W9-BZW-467

The thermodynamic Universe

Exploring the Limits of Physics

The thermodynamic Universe

Exploring the Limits of Physics

B G Sidharth

International Institute of Applicable Mathematics
and Information Sciences, Hyderabad, India

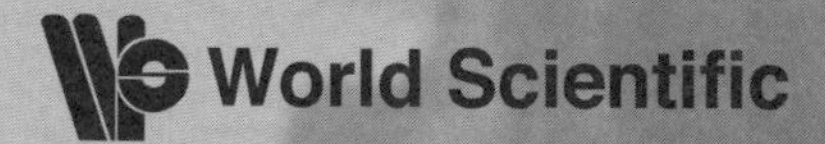

NEW JERSEY • LONDON • SINGAPORE • BEIJING • SHANGHAI • HONG KONG • TAIPEI • CHENNAI

Published by

World Scientific Publishing Co. Pte. Ltd.
5 Toh Tuck Link, Singapore 596224
USA office: 27 Warren Street, Suite 401-402, Hackensack, NJ 07601
UK office: 57 Shelton Street, Covent Garden, London WC2H 9HE

British Library Cataloguing-in-Publication Data
A catalogue record for this book is available from the British Library.

THE "THERMODYNAMIC" UNIVERSE
Exploring the Limits of Physics

ISBN-13 978-981-281-234-6
ISBN-10 981-281-234-2

Printed in Singapore by B & JO Enterprise

Preface

The progress of Science has had its share of twists and turns, whether it be at the beginning of the nineteenth century or the beginning of the twentieth century, in more recent times. On these two occasions new ideas had to be invoked, indeed were forced upon us which led to a paradigm shift−−the Atomic Theory and its consequences in the nineteenth century and Relativity and Quantum Theory in the twentieth century. After a century of dedicated and even frenetic work by some of the greatest minds in the world, it is now becoming apparent that we have reached yet another such turning point.

The attempts to provide a unified description of Quantum Theory and General Relativity or Gravitation have led to approaches like the String Theory and Derivative Theories. There has been impressive progress, particularly in String Theories over the past few decades. At the cosmic scale, the Standard Big Bang Cosmology had been perfected. A complete description of the Universe seemed to be falling into place.

This rosy picture was spoilt in the last years of the twentieth century. On the one hand, we were suddenly confronted with the realization that far from a decelerating Universe, being held back by Dark Matter, the Universe is actually accelerating, driven by Dark Energy. Indeed this had been predicted by the author in 1997 itself. At the micro scale, slowly the realization dawned that the various String Theories were leading to more and more exotic but unrealistic scenarios−−our expectation of a Theory of Everything remained far from reality.

We would like to suggest that we are confronted with yet another paradigm shift. Rather than the reductionist approach which was at the heart of twentieth century physics, perhaps we have to invoke a Universe that is "Thermodynamic" in nature, in that fundamental properties result from collective or cooperative phenomena. That is the theme of this book, which is based on some hundred papers written by the author over the past decade as also two books, "The Chaotic Universe: From the Planck to the Hubble Scale" published by Nova Science, New York and "The Universe of Fluctuations" published by Springer.

There are some satisfying features. Firstly, there is contact and agreement with experiment and observation. Indeed as Prof. Abdus Salam would say, "experiment is at the heart of physics". Secondly, established conventional theories follow as suitable limiting cases of the present ideas. Furthermore, the ideas are simple and explain several phenomena at once whether it be a mass spectrum formula that gives the masses of all known elementary particles (and predicts others) or the deduction from theory of the supposedly miraculous, but empirical Large Number relations--and so on. This is very much in the spirit of science, finding a common denominator for the goings on in the Universe.

Following the suggestion of some important reviewers, the book has been written with a pedagogic flavour so as to be accessible to a larger audience of graduate and informed students. This has been done at the risk of being repetitive on the one hand, and on the other several simple references have been given.

I would particularly like to express my thanks to my friend, Prof. Walter Greiner, Director of the Frankfurt Institute for Advanced Studies, for his encouragement and suggestions and Mrs. Y. Padma for painstakingly preparing the manuscript.

B.G. Sidharth
B.M.Birla Science Centre
Hyderabad
December 2007

Contents

Chapter 1

The Limits of Physics

1.1 Our Scientific Legacy

From time immemorial, human beings have observed the universe around them and have tried to understand and explain the phenomena they saw. In the process they built models, that is described new events in terms of concepts they already knew. This has been an ongoing process till date.
The earliest known model builders of the universe were the composers of the hymns of the Rig Veda, some ten thousand years ago. With amazing insights, they described the Earth and sky as two bowls [1]. They went on to describe the Sun as a star of the daytime sky and even asked, how is it that though the Sun is not bound it does not fall down? And so on and so on. For the ancient Egyptians of four to five thousand years ago, the sky was supported, at the extreme ends by mountains.
Perhaps the earliest model of what we today call microphysics was proposed by the ancient Indian thinker Kanada who lived around the seventh century B.C. For him the universe was made up of ultimate sub constituents which were in perpetual vibration [2]. Later Greeks also had an Atomic Theory, which they may or may not have acquired from India. But there was a crucial difference. Their atoms were static.
Our legacy of modern science came from these Greeks who built up over a few centuries, an even more complex cosmic scheme in which the Earth was at the centre, surrounded by a series of transparent material spheres to which the various heavenly objects like the Sun, Moon, planets and stars were attached. The material spheres were necessary, for, otherwise they would have had to explain why the Moon doesn't crash down on to the Earth, for example. These were spheres because Plato had preached that the circle (or sphere) was a perfect object, due to its total symmetry.

Furthermore these spheres would be in rotation to explain the drama of heavenly motions.
As the observations became more and more precise, the above simple model, first put forward by Anaximenes around 500 B.C. needed modifications [3]. For instance the centre of a sphere would not coincide exactly with the Earth, but rather would be eccentric, that is, slightly away from it. Then the spheres themselves had to carry additional spheres called epicycles, themselves spinning and the objects were placed on top of the epicycles. Ptolemy the Librarian of Alexandria compiled all this knowledge in two astronomical treatises, only one of which, the Great Astronomer or Al Megast survived. The Ptolemaic universe was a complicated tangle of such spheres and epicycles, undergoing complex circular motions.
These basic ideas survived for nearly two thousand years, till the time of Kepler, in fact. Early in the seventeenth century, Kepler noticed that the Greek model differed from observation by just eight minutes of arc, for the orbit of Mars. Kepler had inherited the meticulous observations of Tycho Brahe, and a lesser mortal would have attributed this minor discrepancy to an error in observation. On the contrary, Kepler was convinced that the observations were correct and that the discrepancy pointed to a reformation of Astronomy. Clearly the limit of the validity of the Greek model had been reached.
Kepler proposed his first two laws of planetary motion around 1608. Some years later the third law followed. Crucially the orbits were ellipses. With a single ellipse Kepler could explain the minute discrepancy between theory and observation, for the planet Mars. What the Greeks had tried to do was, approximate a simple elliptical motion by a series of complicated circular motions. The larger implication of this minute correction was this: The ellipse destroyed the crystal spheres of the Greek model and the age old question was once again thrown open: Why dont the Moon, the planets and so on crash down?
This question was answered by Newton who needed the laws of mechanics which had been developed a little earlier by Galileo. He introduced his Theory of Gravitation. Kepler's purely observational laws could now be explained from theory.
Newtonian Mechanics dominated the scientific scene for a few centuries. There was an absolute space, while time was separate and reversible. The equations of mechanics were valid if the time t were replaced by $-t$. Another important concept implicit in Newtonian Mechanics and gravitation theory was action at a distance. Every object exerted instantaneously a

gravitational force on every other object.
However, in the nineteenth century, a new discipline was born, which also had a new ethos−−rather than being an abstract study of the universe, this new discipline, Thermodynamics was a child of the industrial era. In the words of Toffler [4]:
"In the world model contributed by Newton and his followers, time was an after thought. A moment whether in the present, past, or future, was assumed to be exactly like any other moment...
"In the nineteenth century, however, as the main focus of physics shifted from dynamics to thermodynamics and the Second Law of Thermodynamics was proclaimed, time suddenly became a central concern. For, according to the Second Law, there is an inescapable loss of energy in the universe. And, if the world machine is really running down and approaching the heat death, then it follows that one moment is no longer exactly like the last. You cannot run the universe backward to make up for entropy. Events over the long term cannot replay themselves. And this means that there is a directionality or, as Eddington later called it, an "arrow" in time. The whole universe is, in fact, aging. And, in turn, if this is true, time is a one-way street. It is no longer reversible, but irreversible.
"In short, with the rise of thermodynamics, science split down the middle with respect to time. Worse yet, even those who saw time as irreversible soon also split into two camps. After all, as energy leaked out of the system, its ability to sustain organized structures weakened, and these, in turn, broke down into less organized, hence more random elements. But it is precisely organization that gives any system internal diversity. Hence, as entropy drained the system of energy, it also reduced the differences in it. Thus the second Law pointed toward an increasingly homogeneous−−and, from the human point of view, pessimistic−−future.
"... time makes its appearance with randomness: "Only when a system behaves in a sufficiently random way may the difference between past and future, and therefore irreversibility, enter its description." In classical or mechanistic science, events begin with "initial conditions," and their atoms or particles follow "world lines" or trajectories. These can be traced either backward into the past or forward into the future. This is just the opposite of certain chemical reactions, for example, in which two liquids poured into the same pot diffuse until the mixture is uniform or homogeneous. These liquids do not de-diffuse themselves. At each moment of time the mixture is different, the entire process is "time-oriented."
In a sense these "thermodynamic" ideas were anticipated in the nineteenth

century itself, through the work of Poincare and others, working in the field of celestial mechanics rather than industrial machines. Were the orbits of the planets or other celestial objects really unchanging in time? Poincare realized that celestial mechanics had been worked out under the banner of what may be called the two body problem. The orbit of the earth round the Sun, for example, would be more or less unchanging, if the Earth and the Sun were the only two objects in the universe. Even with a third planet, we have to consider the three body problem, which as Poincare realized had no analytical solution. He had laid the ground for what has subsequently come to be known as the chaos theory. As Prigogine was to say much later [5]:

"Our physical world is no longer symbolized by the stable and periodic planetary motions that are at the heart of classical mechanics. It is a world of instabilities and fluctuations..."

Definitely the limits of Newtonian Mechanics had been reached.

The nineteenth century also saw the birth and development of yet another discipline, Electromagnetism. Now at this stage the earlier action at a distance concept had to be abandoned. Maxwell's work introduced the new paradigm of a field. Earlier an electric charge was conceived of as acting on another charge, via the Coulomb force, very much like Newton's gravitational force. This idea is correct, if the two charges are at relative rest, a situation which does not exist in the real world. When the charges move, more correctly accelerate, the interaction of one charge travels through the intervening medium, in the form of electromagnetic waves, which impinge upon the other charge at a later time, unlike the instantaneous action of Newtonian gravitation.

The stage had now been set for Einstein's Special Theory of Relativity. The Special Theory of Relativity introduced two ideas, one of which appeared to be self contradictory. Nevertheless these two ideas explained the puzzling and indeed otherwise inexplicable consequences of the Michelson-Morley and similar experiments. The point was that light had been thought of as electromagnetic radiation traveling with the same speed. In this case its speed would be different for different observers in relative motion. However the Michelson-Morley experiments showed that this was not so.

Einstein proposed at the turn of the twentieth century that the velocity of light would be the same in all directions−−a relatively easy idea to digest. But then he also had a second postulate−−the speed of light would be the same for all observers, moving with a uniform velocity with respect to one another. How could this be? It blatantly contradicted Newtonian Mechanics. Einstein could show that this was a contradiction if we retained the Newtonian concepts of space and time. If on the other hand, we realized that space and time get mixed up and that the lengths of the intervals of spacetime, which had been taken to be the same for all observers in Newtonian Mechanics, were on the contrary different for different observers, the contradiction would be removed. Clearly the limits of Newtonian Mechanics had been reached yet again.

When Einstein proposed his Special Theory of Relativity, there were two ruling paradigms, which continue to hold sway even today, though not so universally. The first was that of point elementary particles and the second was that of space time as a differentiable manifold. Further, Einstein's work introduced the concept of causality−−no signal could travel faster than light. So, the effect−−gravitational, electromagnetic, whatever−−of one object would be felt by another object at a later time, and not instantaneously, as in the earlier theory. That is, the signals would be retarded. Little wonder therefore that as the relativistic theory of the electron developed, there were immediate inconsistencies which were finally ostensibly resolved only with the intervention of Quantum Theory. This was because, historically the original concept of the electron was that of a spherical charge distribution [6–8]. It is interesting to note that in the non-relativistic case, it was originally shown that the entire inertial mass of the electron equalled its electromagnetic mass. The question came up, was this a meeting between electromagnetism and mechanics? This motivated much work and thought in this interesting direction. To put it briefly, in non relativistic theory, we get [6],

$$\text{Kinetic} \quad \text{energy} = (\beta/2)\frac{e^2}{Rc^2}v^2,$$

where R is the radius of the electron and β is a numerical factor of the order of 1. So we could possibly speak of the entire mass of the electron in terms of its electromagnetic properties.

It might be mentioned that it was possible to think of an electron as a charge distribution over a spherical shell within the relativistic context too, as long as the electron was at rest or was moving with a uniform velocity. However it was necessary to introduce, in addition to the electromagnetic force, the Poincare stresses−−these were required to counter balance the mutual repulsive "explosion" of the different parts of the electron.

When the electron in a field is accelerated, the above picture no longer holds. We have to introduce the concept of the electron self force which is given by, in the simple case of one dimensional motion,

$$F = \frac{2}{3}\frac{e}{Re^2}\ddot{x} - \frac{2}{3}\frac{d}{dt}\ddot{x} + \gamma\frac{e^2 R}{c^4}\ddot{x} + 0(R^2) \tag{1.1}$$

where dots denote derivatives with respect to time, and R as before is the radius of the spherical electron. More generally (1.1) becomes a vector equation. In (1.1), the first term on the right side gives the electromagnetic mass of the earlier theory. As can be seen from (1.1), as R the size of the electron $\to 0$ the first term $\to \infty$ and this is a major inconsistency. It was the first of a series of infinities that has plagued twentieth century physics. In contrast the second term which contains the non Newtonian third time derivative remains unaffected while the third and following terms $\to 0$. It may be mentioned that the first term (which $\to \infty$) gives the electromagnetic mass of the electron while the second term gives the well known Schott term (Cf.ref.[6, 7, 9]). Its presence is required however because it compensates the energy loss due to radiation by the accelerated electron. In any case it is possible to develop a model of an extended electron consistent with relativity on these lines, but at the expense of introducing non electromagnetic forces.
Let us now see how it was possible to rescue the relativistic electron theory, though at the expense of introducing some unphysical concepts.

1.2 The Advanced and Retarded Fields

To proceed, from a classical point of view a charge that is accelerating radiates energy which dampens its motion. This is given by the second term on the right side of (1.1). Dirac proposed in 1938 a phenomenological equation that overcomes the infinite (electromagnetic) mass in (1.1). The Lorentz Dirac equation, which in units $c = 1$, and τ being the proper time, while $\imath = 1, 2, 3, 4$, is (Cf.[10]),

$$m\frac{d^2x^\imath}{d\tau^2} = eF^\imath_k\frac{dx^k}{d\tau} + \frac{4e}{3}g_{lk}\left(\frac{d^3x^\imath}{d\tau^3}\frac{dx^l}{d\tau} - \frac{d^3x^l}{d\tau^3}\frac{dx^\imath}{d\tau}\right)\frac{dx^k}{d\tau}, \tag{1.2}$$

This holds for a point charge, m being a "renormalized mass" that absorbs the infinity. Here is the precursor of renormalization, that has gone hand in hand with the infinities of twentieth century physics. The first term gives the usual external field while the second term does not come from the Lagrangian (which gives the first term and the Lorentz force)—–it comes by putting in energy conservation (due to radiation loss) by hand. Equation

(1.1) can be written as

$$m\frac{d^2x^\imath}{d\tau^2} = e\{F^\imath_k + R^\imath_k\}\frac{dx^k}{d\tau} \tag{1.3}$$

where

$$R^\imath_k \equiv \frac{1}{2}\{F^\imath_{k(ret)} - F^\imath_{k(adv)}\} \tag{1.4}$$

In (1.4), $F_{(ret)}$ denotes the retarded or causal field allowed by relativity, as alluded to. $F_{(adv)}$ on the other hand is the advanced field that is unphysical, in the sense that it is not sanctioned by relativity. While the former is the causal field where the influence of a charge at A is felt by a charge at B at a distance r after a time $t = \frac{r}{c}$, the latter is the advanced field which acts on A from a future time. In effect what Dirac showed was that the radiation damping term in (1.2) or (1.3) is given by (1.4) in which an antisymmetric difference of the advanced and retarded fields is taken. Let us elaborate a little further.

The Maxwell wave equation has two independent solutions, one having support on the future light cone, this is the retarded solution and the other having support on the past light cone which has been called the advanced solution. The retarded solution is selected to describe the physical situation in conventional theory taking into account the usual special relativistic concept of causality. This retarded solution is physically meaningful, as it describes electromagnetic radiation which travels outward from a given charge with the speed of light and reaches another point at a later instant. It has also been called for this reason the causal solution. On the grounds of this causality, the advanced solution has been rejected, except in a few formulations like those of Dirac above, or Feynman and Wheeler (F-W) to be seen below.

It must also be mentioned that Dirac's prescription lead to the so called runaway solutions, with the electron acquiring larger and larger velocities in the absence of an external force [11]. This he related to the infinite self energy of the point electron.

To elaborate further, we use the difference of the advanced and retarded fields in (1.1), in the following manner: We use successively $F_{(ret)}$ and $F_{(adv)}$ in (1.1) and take the difference in which case the self force becomes (Cf.[9])

$$F = -\frac{2}{3}\frac{e^2}{c^3}\frac{d}{dt}(\ddot{x}) + 0(R)$$

In the above, the troublesome infinity generating term of (1.1) is absent, while the third derivative term is retained. On the other hand this term is required on grounds of conservation of energy, due to the fact that an accelerated electron radiates energy (Cf.[12]). Except for the introduction of advanced fields, we have infinity free results. However, in this formulation too, there is no electromagnetic mass term, and further, as will be seen below, we have to extend our considerations to a small neighborhood of the electron, and not just the point electron itself. To see this in detail, we observe that the well known Lorentz-Dirac equation (Cf.[6]), can be written as

$$ma^{\mu}(\tau) = \int_0^{\infty} K^{\mu}(\tau + \alpha\tau_0)e^{-\alpha}d\alpha \tag{1.5}$$

where a^{μ} is the acceleration and

$$K^{\mu}(\tau) = F^{\mu}_{in} + F^{\mu}_{ext} - \frac{1}{c^2}\bar{R}v^{\mu},$$

$$\tau_0 \equiv \frac{2}{3}\frac{e^2}{mc^3} \sim 10^{-23}sec \tag{1.6}$$

and

$$\alpha = \frac{\tau' - \tau}{\tau_0},$$

where τ denotes the time and $\bar{R}$ is the total radiation rate.

It can be seen that equation (1.5) differs from the usual equation of Newtonian Mechanics, in that it is non local in time. That is, the acceleration $a^{\mu}(\tau)$ depends on the force not only at time τ, but at subsequent times also. Let us now try to characterize this non locality. We observe that τ_0 given by equation (1.6) is the Compton time $\sim 10^{-23}secs$. This is the precursor of Quantum Theory. So equation (1.5) can be approximated by

$$ma^{\mu}(\tau) = K^{\mu}(\tau + \xi\tau_0) \approx K^{\mu}(\tau) + \xi\tau_0\dot{K}^{\mu}(\tau) + \cdots \tag{1.7}$$

Thus as can be seen from (1.7), the Lorentz-Dirac equation differs from the usual local theory by a term of the order of

$$\frac{2}{3}\frac{e^2}{c^3}\dot{a}^{\mu} \tag{1.8}$$

the so called Schott term. It is well known that the time component of the Schott term (1.8) is given by (Cf.ref.[6])

$$-\frac{dE}{dt} \approx \bar{R} \approx \frac{2}{3}\frac{e^2c}{r^2}\left(\frac{E}{mc^2}\right)^4,$$

where E is the energy of the particle. Whence integrating over the period of non locality $\sim \tau_0$ the Compton time, we can immediately deduce that r the scale of spatial non locality is given by

$$r \sim c\tau_0,$$

which is of the order of the Compton wavelength as indeed can be expected. So far as the breakdown of causality is concerned, this takes place within a period $\sim \tau$, the Compton time as we briefly saw [6, 11].

In the F-W formulation on the other hand, the rest of the charges in the universe react back on the original electron through their advanced waves, which arrive (from the future) at the given charge at the same time as the given charge radiates its electromagnetic waves. More specifically, when an electron is accelerated at the instant t, it interacts with the other charges at a later time $t' = t + r/c$ where r is the distance of the other charge—–these are the retarded interactions. However the other charges react back on the original electron through their advanced waves, which will arrive at the time $t' - r/c = t$. Effectively, there is instantaneous action at a distance. It must be mentioned that in the F-W formulation there is no self force (and therefore the electromagnetic mass and the infinite term—–the first term on the right side of (1.1)) or radiation damping. This is provided instead by the action of all other charges in the universe on the original charge.

Let us throw further light on all this. There are two important inputs which we can see in the above formulation. The first is the action of the rest of the universe at a given charge and the other is spacetime intervals which are of the order of the Compton scale. In fact we can push the above calculations further. The work done on a charge e at O by the charge P at a distance r in causing a displacement $x \sim l$ is given by

$$\frac{e^2 l}{r^2}$$

Now the number of particles at distance r from O is given by

$$n(r) = \rho(r) \cdot 4\pi r^2 dr$$

where $\rho(r)$ is the density of particles. So the total work is given by

$$E = \int \int \frac{e^2}{r^2} l 4\pi r^2$$

which can be shown to be $\sim mc^2$. This is because,

$$\rho(r) = N/R^3,$$

where N is the total number of particles in the universe, R now is its radius and anticipating a result from Chapters 2 and 3,

$$R \sim \sqrt{N}l,$$

where l is given by (1.6).
Wheeler and Feynman thus reformulated the above action at a distance formalism in terms of what has been called their Absorber Theory. In their formulation, the field that a charge would experience because of its action at a distance on the other charges of the universe, which in turn would act back on the original charge is given by

$$\bar{R}e = \frac{2e^2 d}{3dt}(\ddot{x}) \tag{1.9}$$

The interesting point is that instead of considering the above force in (1.9) at the charge e, if we consider the response at an arbitrary point in its neighborhood as was shown by Feynman and Wheeler (Cf.ref.[13]) and, in fact a neighborhood at the Compton scale, as we saw above and was argued by the author [14], the field would be precisely the Dirac field given in (1.3) and (1.4).
The net force emanating from the charge is thus given by

$$F^{ret} = \frac{1}{2}\left\{F^{ret} + F^{adv}\right\} + \frac{1}{2}\left\{F^{ret} - F^{adv}\right\} \tag{1.10}$$

which is the acceptable causal retarded field. The causal field now consists of the time symmetric field which implies no radiation of the charge together with the Dirac field, that is the second term in (1.10), which now represents the response of the rest of the charges. Interestingly in this formulation we have used a time symmetric field, viz., the first term of (1.10) to recover the retarded field with the correct arrow of time.
Feynman and Wheeler stressed that the universe has to be a perfect absorber or to put it simply, every charged particle in the universe should respond back to the action on it by the given charge in our instantaneous action at a distance scenario. In the Feynman-Wheeler formulation to reiterate there is no electromagnetic mass and also no radiation damping——we have finally the retarded field; but within the context of the Instantaneous Action at a Distance. In any case, it was realized that the limits of classical physics are reached in the above considerations, at the Compton scale. However, we will now argue that there is actually a convergence between Classical and Quantum Physics.
There are two important inputs which we would like to re-emphasize in the above more recent formulation. The first is the action of the rest of the

universe at a given charge and the other is minimum spacetime intervals which are of the order of the Compton scale. The minimum spacetime interval removes, firstly the advanced field effects which take place within the Compton time and secondly the infinite self energy of the point electron disappears due to the Compton scale. We thus bypass renormalization. This would be an important idea in the rest of the book.

1.3 Quantum Mechanical Considerations

The Compton scale comes as a Quantum Mechanical effect, within which we have zitterbewegung effects and a breakdown of causal physics [15]. Indeed Dirac had noted this aspect in connection with two difficulties with his electron equation. Firstly the speed of the relativistic Quantum Mechanical electron turns out to be the velocity of light. Strictly speaking, this would imply an infinite mass for the electron. Secondly the position coordinates become complex or non Hermitian. This is physically meaningless. His explanation was that in Quantum Theory we cannot go down to arbitrarily small spacetime intervals, for the Heisenberg Uncertainty Principle would then imply arbitrarily large momenta and energies. So Quantum Mechanical measurements are an average over intervals of the order of the Compton scale. Once this is done, we recover meaningful physics. All this has been studied afresh by the author more recently, in the context of a fuzzy non differentiable spacetime and noncommutative geometry [16]. This indeed, will be the theme of this book. We will first argue that there is a convergence between preceding considerations and Quantum Mechanical Theory. The Compton scale that surfaces in both these considerations, already gives a hint of this.

Weinberg too notices the non physical aspect of the Compton scale [17]. Starting with the usual light cone of Special Relativity and the inversion of the time order of events, he goes on to add, "Although the relativity of temporal order raises no problems for classical physics, it plays a profound role in quantum theories. The uncertainty principle tells us that when we specify that a particle is at position x_1 at time t_1, we cannot also define its velocity precisely. In consequence there is a certain chance of a particle getting from x_1 to x_2 even if $x_1 - x_2$ is space-like, that is, $|x_1 - x_2| > |x_1^0 - x_2^0|$. To be more precise, the probability of a particle reaching x_2 if it starts at

x_1 is nonnegligible as long as

$$0 \leq (x_1 - x_2)^2 - (x_1^0 - x_2^0)^2 \leq \frac{\hbar^2}{m^2} \cdots \quad (1.11)$$

where $\hbar$ is Planck's constant (divided by 2π) and m is the particle mass. (Such space-time intervals are very small even for elementary particle masses; for instance, if m is the mass of a proton then $\hbar/m = 2 \times 10^{-14}cm$ or in time units $6 \times 10^{-25}sec$. Recall that in our units $1sec = 3 \times 10^{10}cm$.) We are thus faced again with our paradox; if one observer sees a particle emitted at x_1, and absorbed at x_2, and if $(x_1 - x_2)^2 - (x_1^0 - x_2^0)^2$ is positive (but less than or $= \hbar^2/m^2$), then a second observer may see the particle absorbed at x_2 at a time t_2 before the time t_1 it is emitted at x_1.
"There is only one known way out of this paradox. The second observer must see a particle emitted at x_2 and absorbed at x_1. But in general the particle seen by the second observer will then necessarily be different from that seen by the first."
There is another way to view (1.11). The light cone of special relativity viz., $(x_1 - x_2)^2 - (x_1^0 - x_2^0)^2 = 0$ now gets somewhat distorted because of Quantum Mechanical effects.
Let us now consider the above in the context of a non zero photon mass. Such a mass $\sim 10^{-65}gms$ was rather recently deduced by the author, and it is not only consistent with experimental restrictions, but also predicts a new effect viz., a residual cosmic radiation $\sim 10^{-33}eV$, which in fact has been observed [18–22]. We will come back to this in detail in later Chapters, particularly Chapter 4. Such a photon would have a Compton length $\sim 10^{28}cms$, that is the radius of the universe itself.
This would then lead to the following scenario: An observer would see a photon leaving a particle A and then reaching another particle B, while a different observer would see exactly the opposite for the same event—–that is a photon leaves B and travels "backward" in time to A, as in the Weinberg interpretation. This latter gives the advanced potential. We are back with the Feynman-Wheeler instantaneous action scenario. The distinction between the advanced and retarded potentials of the old electromagnetic theory thus gets mixed up and we have to consider both the advanced and retarded potentials [13]. Thus, two charged particles interacting via the exchange of photons will be described as above, using (1.11). Indeed in Quantum Field Theory this is described as the exchange of virtual photons.
We consider this in a little more detail: The advanced and retarded solutions of the wave equation are given by the well known advanced and

retarded potentials given by, in the usual notation, the well known expression

$$A^{\mu}_{ret(adv)}(x) = \frac{1}{c}\int \frac{j^{\mu}(x')}{|r-r'|}\delta\left(|r-r'| \mp c(t-t')\right) d^4x'$$

(The retarded part of which leads to the Lienard Wiechart potential of earlier theory).

It can be seen in the above that we have the situation described within the Compton wavelength, wherein there are two equivalent descriptions of the same event—–a photon leaving the charge A and reaching the charge B or the photon leaving the charge B and reaching the charge A. The above expression for the advanced and retarded potentials immediately leads to the advanced and retarded fields (1.4) and (1.10) of the F-W description except that we now have a rationale for this formulation in terms of the photon mass and the photon compton wavelength rather than the perfect absorber ad hoc prescription. In fact there is now an immediate Quantum Mechanical explanation in this of the Instantaneous Action At a Distance Theory alluded to. Thus these considerations reconcile the Quantum Mechanical and Classical pictures. We note however that as the photon mass is so small, the usual theory is still a good approximation.

To sum up [13], the Feynman Wheeler Perfect Absorber Theory required that every charge should interact instantaneously with every other charge in the universe, that is that the universe must be a perfect absorber of all electromagnetic fields emanating from within. If this condition were satisfied, then the nett response of all charged particles along the future light cone of the given charge is expressed by an integral that converges. We have argued that this ad hoc prescription of Feynman and Wheeler as embodied by the inclusion of the advanced potential is automatically satisfied if we consider the photon to have a small mass $10^{-65}gms$ which is consistent with the latest experimental limits—–this leading to the effect mentioned by Weinberg within the Compton wavelength, which is really the inclusion of the advanced field as well.

To put the above in different words, when we talk of two (charged) particles A and B and the instant t, we are attributing the same t to A and the distant B. This is consistent with Special Relativity. This enables us to talk of an advanced wave leaving B at $t + \Delta t$ and travelling "backward" in time to reach A at t. This simultaneity however, breaks down within the Quantum Mechanical Compton time. We could very well describe the event as an ordinary retarded wave leaving B at $t - \Delta t$ and reaching A at t.

1.4 The Limits of Special Relativity

What we have witnessed above is that it is still possible to rescue the classical relativistic theory of the electron, but at the expense of introducing the advanced fields into the physics, fields which have been considered to be unphysical.
Another perspective is, as seen above, that there is instantaneous action at a distance, which apparently goes against relativistic causality. But let us now note that in both the Dirac and the Feynman-Wheeler approaches, we are no longer dealing with point particles alone, but rather with a small neighborhood of such a point particle, a neighborhood of a Compton length dimension. Furthermore within the Compton scale, relativistic causality breaks down as embodied in (1.11).
We can then reformulate the above considerations in the following manner: The limit of applicability or the limit of validity of the relativistic electron theory as also the Special Theory of Relativity is the Compton scale of a particle. The points within the Compton scale no longer obey Special Relativity and see a non relativistic, instantaneous action at a distance universe. Indeed Rohrlich notes [23], "... the notion of a "classical point charge" is an oxymoron because "classical" and "point" contradict one another: Classical physics ceases to be valid at sizes at or below the Compton wavelength and thus cannot possibly be valid for a point object..."

1.5 Discussion

Let us sum up the foregoing considerations. In Classical Physics the point electron leads to infinite self energy via the electromagnetic mass term e^2/R, where R is the radius which is made to tend to zero. If on the other hand R does not vanish, in other words we have an extended electron, then we have to introduce non electromagnetic forces like the Poincare stresses for the stability of this extended object, though on the positive side this allows the radiation damping or self force that is required by conservation laws.
Dirac could get rid of these problems by introducing the difference between the advanced and retarded potentials in his phenemenological equation in which the infinity was absorbed into a renormalized point particle mass: This was the forerunner of the renormalization theory and was the content of the Lorentz-Dirac equation. The new term represents the radiation damping effect, but we then have to contend with the advanced potential

or equivalently a non locality in time. However this non locality takes place within the Compton time, within which the electron attains a luminal velocity.

The Lorentz-Dirac equation also had unsatisfactory features like the non-Newtonian derivative of the acceleration, the non locality in time and the run away solutions, features confined to the Compton scale.

The Feynman-Wheeler approach bypasses the infinity and the extended electron self force−−but the mass is no longer electromagnetic. Moreover the nett result is that there is only the desired retarded potential, but an instantaneous interaction with the rest of the charges of the universe has to be invoked. It is this interaction with the remaining charges which leads to the point electron's self energy. Surprisingly however the interaction with the rest of the charges in the immediate vicinity of the given charge in the Feynman-Wheeler formula gives us back the Dirac antisymmetric difference with its non locality within the Compton scale. There is thus a reconciliation of the Dirac and the Feynman Wheeler approaches, once we bring into the picture, the Compton scale.

Outside this scale, however, the theory is causal that is uses only the retarded potential because effectively the advanced potential gets canceled out as it appears as the sum of the symmetric and antisymmetric differences.

The final conclusion was that in a classical context a totally electromagnetic electron is impossible as also the concept of a point electron without introducing additional "unphysical" concepts including action at a distance. It was believed therefore that the electron was strictly speaking the subject of Quantum Theory.

Nevertheless in Dirac's relativistic Quantum Electron, we again encounter the electron with the luminal velocity within the Compton scale, precisely what was encountered in Classical Theory as well, as noted above. This again is the feature of a point space time approach. At this stage a new input was given by Dirac−−meaningful physics required averages over the Compton scale, in which process, the unphysical zitterbewegung effects were eliminated. Nor has Quantum Field Theory solved the problem−−one has to take recourse to renormalization, and as pointed out by Rohrlich, one still has a non electromagnetic electron. In any case, it appears that further progress would come either from giving up point spacetime or from an electron that is extended (or has a sub structure) in some sense [8, 7, 11, 6]. From this point of view the relativistic theory of the electron is inconclusive to date. As noted by Feynman himself in his famous Lectures on Physics

(Vol II), "We do not know how to make a consistent theory−−including the Quantum Mechanics−−which does not produce an infinity for the self energy of the electron, or any point charge. And at the same time there is no satisfactory theory that describes a non-point charge..."
In the words of Hoyle and Narlikar [10], "...it was believed that the problem of the self force of the charge would not be solved except by recourse to Quantum Theory... This hope has not been fully realized. Quantum Field Theory does alleviate the self energy problem but cannot surmount it without introducing the renormalization programme..."
Indeed Dirac himself was unhappy with renormalization, which he termed an accident. He expressed his confidence that it would be disproved eventually.
In his words, "I am inclined to suspect that the renormalization theory is something that will not survive in the future, and that the remarkable agreement between its results and experiments should be looked on as a fluke..."
We have pointed out that the important point however is that all this can be explained consistently in Quantum Mechanical terms in the context of the photon having a non zero mass, consistent with experiment $\sim 10^{-65} gms$.
So there is convergence between the Dirac and the Feynman-Wheeler approaches if we consider the fact that special relativity, as seen above, does not hold within the Compton wavelength. This explains the non locality in time. This justifies the use of the advanced potential or non locality in time of the Lorentz-Dirac approach or also the fact that a point inside the Compton wavelength sees a non relativistic instantaneous action at a distance universe around it−−this is the instantaneous action at a distance of the Feynman-Wheeler approach. Furthermore, the radiation of photons emitted by the accelerated electrons (in the Dirac self force) are meaningful only if they impinge on other charges as in the Field Theory.
We now briefly re-emphasize the following.
1. In classical relativistic theory, there appeared an impasse. We could get a special relativistic electron with cohesive forces in an extended model but at the expense of the purely electromagnetic electron. On the other hand point electrons were not meaningful as their self energy diverged. Consequently the structure dependent terms for example in (1.1) had to be taken seriously.
2. We have arrived at the Compton scale from two different approaches. Classically, there was the electron radius and Quantum Mechanically the Compton length, both of the same order except for a factor of the order of

the fine structure constant:

$$\hbar/mc \sim \beta \cdot e^2/mc^2$$

The left side has the Quantum Mechanical Planck constant while the right side has merely classical quantities. We could consider this to be a derivation of the rough value of the Planck constant of Quantum Mechanics, in an order of magnitude sense. We will return to this point in a later Chapter.
3. In any case the above considerations at the Compton scale lead in recent studies to a noncommutative geometry and the limit to a point particle no longer becomes legitimate. This will be extensively discussed in this book.

1.6 The Quantum Universe

The advent of Quantum Mechanics threw up several, what may be called counter intuitive ideas and even Einstein could not reconcile to them. One of these ideas was the wave particle duality. Another was Heisenberg's Uncertainty Principle: surprisingly it would not be possible to measure simultaneously and accurately the position and momentum of a particle. This was related to wave particle duality itself. Yet another was that of the collapse of the wave function in which process causality becomes a casuality. To put it simply, if the wave function is a super position of the eigen states of an observable, then a measurement of the observable yields one of the eigen values no doubt, but it is not possible to predict which one. Due to the act of observation, the wave function instantly collapses to any one of its eigen states in an acausal manner. To put it another way, the wave function obeys the causal Schrodinger equation, for example, till the instant of observation at which point, causality ceases. Indeed, we saw that this was true within the Compton scale itself.
Another important counter intuitive feature of Quantum Mechanics is that of non locality. In fact Einstein with Podolsky and Rosen put forward in 1935 his arguments for the incompleteness of Quantum Mechanics on this score [24, 25]. This has later come to be known as the EPR paradox. To put it in a simple way, without sacrificing the essential concepts, let us consider two elementary particles, for example two protons kept together somehow. They are then released and move in opposite directions. When the first proton reaches the point A its momentum is measured and turns out to be say, $\vec{p}$. At that instant we can immediately conclude, without any further measurement that the momentum of the second proton which is at

the point B is $-\vec{p}$. This follows from the Conservation of Linear Momentum, and is perfectly acceptable in Classical Physics, in which the particles possess a definite momentum at each instant.
In Quantum Physics, the difficulty is that we cannot know the momentum at B until and after a measurement is actually performed, and then that value of the momentum is unpredictable. What the above experiment demonstrates is that the proton at B instantly came to have the value $-\vec{p}$ for its momentum without any further measurement, when the momentum of the proton at A was measured. This "instant" or "spooky action at a distance" feature was unacceptable to Einstein.
In Quantum Theory however this is legitimate because of another counter intuitive feature which is called Quantum Non-separability. That is, if two systems interact and then separate to a distance, they still have a common state vector. This goes against the concept of locality and causality, because it implies instantaneous interaction between distant systems. So in the above example, even though the protons at A and B may be separated, they still have a common wave function which collapses to some value with the measurement of the momentum of any one of them and self-consistently provides an explanation of the fact that the momentum of the other particle is automatically known without requiring another measurement. This non-separability has been characterized by Schrodinger in the following way: "I would not call that *one*, but rather *the* characteristic of Quantum Mechanics." For Einstein however this was like spooky action at a distance. All this has been experimentally verified since 1980 which sets at rest Einstein's objections.
However this "entanglement" as it is called these days, between distant objects in the universe, does not really manifest itself though it is perfectly legitimate and observable in a universe that consists of let us say just two particles. But a measurement destroys the entanglement. Now in the universe at large as there are so many particles and correspondingly a huge amount of interference, the entanglement is considerably weakened. This was the crux of Schrodinger's arguments. What is these days called decoherence works along these lines. This is in fact the explanation for the famous "Schrodinger's Cat" paradox.
This paradox can be explained in the following simple terms: A cat is in an enclosure along with, let us say a microscopic amount of radioactive material. If this material decays, emitting let us say an electron, the electron would fall on a vial of cyanide, releasing it and killing the cat in the process. Let us say that there is a certain probability of such an electron

being emitted. So there is the same probability for the cat to be killed. There is also a probability that the electron is not emitted, so that there is the same probability for the cat to remain alive. The cat is therefore in a state which is a superposition of the alive and dead states. It is only when an observer makes an observation that this superposed wave function collapses into either the dead cat state or the alive and kicking cat state, and this happening is acausal. So it is only on an observation being made that the cat is killed or saved, and that too in an unpredictable manner. Till the observation is made the cat is described by the superposed wave function and is thus neither alive nor dead.

The resolution of this paradox−−it is a paradox−−is of course quite simple. The paradox is valid if the system consists of such few particles and at such distances that they do not interact with each other. Clearly in the real world this idealization is not possible. There are far too many particles and interferences taking place all the time and the superposed wave function would have collapsed almost instantly. This role of the environment has come to be called de-coherence. We will return to this point shortly.

The important point is that all of Classical and Quantum Physics is based on such idealized laws as if there were no interferences present, that is what we have called a two body scenario, is implicit. Clearly this is not a real life scenario.

1.7 The Strong and Weak Interactions

A major achievement of the twentieth century has been the incorporation of three of the four fundamental interactions, viz., electromagnetism, weak interactions and the strong interaction within a unified mathematical framework. This framework is the non Abelian gauge field theory [26–31]. Though the three forces remain different, the underlying mechanism is the same. From this point of view they could be thought to be different aspects of a single underlying process.

Thus there are leptons and there are quarks. The difference between these sets of particles which are perceived today arise because the Universe has become cold. At sufficiently high energies $\sim 10^{15} GeV$, leptons and quarks would be interchangeable and so also all the three forces would have the same strength. It must be mentioned that the above energy is still beyond the reach of foreseeable accelerators.

Apart from leptons and quarks, which are Fermions, or "material" particles, the fields are mediated by Bosons. These are the photons for electromagnetism, the W and Z Bosons for weak interactions and the gluons for the strong interactions.

Quarks were conceived following the work of Gellmann, Ne'eman and Zweig in the sixties. The motivation had been the overabundance of resonances observed in hadron or strong interaction collisions. These resonances could be classified on the one hand according to the Regge trajectories that plot the angular momentum J versus the mass squared, M^2 [32]. We will touch upon this briefly again. On the other hand, there was the SU(3) classification scheme which related particles of the same spin but different quantum numbers by introducing elemental entities—the quarks—whose combinations could account for all observed hadrons.

It is now believed that there are six kinds of quarks: The down (d), the up (u), the strange (s), the charmed (c), the bottom (b) and the top (t). We attribute to the quarks three colours, red, green and blue which are generalizations of the positive and negative charges. It is these colours which characterize strong interaction and hence this field has come to be known as Quantum Chromo Dynamics (QCD). It may be observed that the leptons do not have any colour and so they do not participate in the strong interactions.

A peculiarity of quarks is their fractional charge—they have either the charge $\frac{1}{3}$ or the charge $\frac{2}{3}$ with their corresponding anti particles having opposite charges. So quarks can combine in two different ways to form hadrons, that is particles like protons and neutrons: Either as quark antiquark pairs or as a triplet of quarks, such that the total charge is either one or zero.

In electromagnetism, or Quantum Electro Dynamics (QED), two charged particles interact by the exchange of a photon, more correctly a virtual photon [33] as noted earlier. This exchange takes place within the Heisenberg Uncertainty time. There is a conservation of electric charge in the process. This combined with the masslessness of the photon is characteristic of the $U(1)$ Group which characterizes QED.

QCD is modelled on QED. However QCD which is described by the SU(3) group is more complicated because it describes interactions of three different colours, unlike QED which deals with just one charge. In QCD the interaction between different colours is expressed in terms of eight massless particles, the gluons, unlike the single photon of QED. Another profound difference is that the gluons do carry colour unlike the photon which is chargeless. The nett result of all this is that there is an effect opposite

to that encountered in the charge screening of QED. In this latter case, an electron is surrounded by virtual electron-positron pairs. The electron attracts the positrons and repels the electrons of these pairs with the result that at large enough distances, the electron charge is shielded by the positrons and so appears reduced. In QCD on the other hand, virtual gluon pairs, themselves carrying colour are formed around a quark, no doubt. But there is now an anti screening effect as if the red component of a gluon is attracted to the red of a quark, for example. So at relatively larger distances, the colour charge of a quark increases and again contrary to the QED scenario, decreases as we approach the quark. The QCD force can therefore be compared to rubber bands—–as we stretch, the elastic force manifests itself, but if the bands slacken at close range, the force decreases and even disappears. It is as if there is confinement at large distances and freedom at shorter, asymptotic distances.
The QCD potential can be written as [30, 34]

$$V(r) = -\frac{\alpha(r)}{r} + \frac{r}{\beta^2}$$

This consists of the Coulombic part $\propto -\frac{1}{r}$ and a confining part $\propto r$. Because of this latter, which dominates for large r, free quarks cannot be observed in nature. After all, the model should explain this fact! On the other hand, the Coulombic part ensures that for small r, the inter quark force vanishes, a circumstance which is called asymptotic freedom. Professors Wilczek, Politzer and Gross were awarded the 2004 Nobel Prize in Physics for this work, done thirty years earlier.
The neutrinos are closely associated with the weak interactions. Though the neutrinos are leptons, they differ from their counterparts in that they are massless (or more precisely, as later discovered, they have a very tiny mass). A massless Fermion exhibits handedness, that is, its spin is either aligned in the direction of its motion (righthanded) or it is aligned anti parallel to its motion (lefthanded). This extra property of handedness characterizes the weak force which violates parity, unlike the other forces (though even the quarks exhibit handedness!). Only lefthanded particles and righthanded anti particles bear a weak charge while the righthanded particles and the lefthanded anti particles are neutral from the point of view of the weak interaction. This interaction acts on doublets of particles, which latter are described by the SU(2) Group, in which particles of a doublet pair can be transformed into one another. The weak interactions are mediated by the W Bosons. However a suitable mixture yields both the photon of electromagnetism and the Z° characterizing weak interactions.

This theory therefore combines electromagnetic and weak interactions and is incorporated in the SU(2) XU(1) group [35].
An important difference between the weak forces on the one hand and QED and QCD on the other is that the intermediate particles of the weak interactions, the W and Z Bosons are not massless, but rather have large masses $\sim 100GeV$. This is characteristic of the fact that the weak charge is not invariably conserved and moreover has an extremely short range $\sim 10^{-15}cms$. We will return to this point later.
One of the problems that has plagued modern field theories is that of infinities. Indeed as we saw, this problem was encountered early in the twentieth century itself when an attempt was made to model the electron as a tiny sphere. If the radius of the sphere was then made to shrink indefinitely, the energy of the electron increased without limit as noted [6]. In QED for instance, if we approach the electron through the shield of screening positrons, the bare charge of the electron would be infinite. It is only the physically observable charge, at a distance, screened by the positron charges, which is finite. It is as if the infinite bare negative charge has been cancelled or neutralized by the infinite screening positive charge, the nett result being the observed finite physical charge. Loosely speaking this procedure is called "renormalization".
Mathematically, we encounter divergent integrals [36]. The infinities are eliminated in two steps. In the first step, called regularization, we introduce constraints, for example a cut off (or a lattice structure), to get a finite result dependent on the regularization parameter like the cut off. Counter terms (dependent on these parameters) are then added to the Lagrangian, such that they cancel the parameter dependent integrals. This generally leads to a rescaling of the mass, charge etc. This is the process that is called Renormalization.
The concept of Renormalization is unsatisfactory from the logical point of view as well as from the point of view of internal consistency. It has provoked unease among Physicists such as Dirac quoted earlier [37] or as we will see 't Hooft and several others. Its merit however, has been that phenomenologically speaking, it works.

1.8 Gauge Fields

It has now come to be recognized that the physical principle governing the fundamental interactions between the elementary particles is gauge invari-

ance. This principle, as we shall see in greater detail later, was originally introduced by Hermann Weyl, though in a different form and with a different motivation viz., the attempt to give a unified General Relativistic description of Electromagnetism and Gravitation [38]. At that time these were the only two known interactions and electrons and protons were the only known elementary particles. Weyl's original theory was soon dismissed as adhoc. But nevertheless it was recognized that gauge invariance was a symmetry of Maxwell's equations with useful implications.
Then in the 1950s Yang and Mills (and Shaw) tried to extend gauge symmetry to other interactions. It must be emphasized that both in Special Relativity and General Relativity there are no absolute frames of reference in the Universe. The physics within a system is independent of the choice of the reference frame. However in Special Relativity this freedom of choice of reference frame is a global symmetry- the Lorentz symmetry. In General Relativity on the other hand, the reference frame is to be defined locally, that is at each and every point in the gravitational field. There are the connections−−the affine connections or Christofell symbols which relate nearby frames in General Relativity, something which is not required in Special Relativity [39].
Weyl attempted to investigate if there were similar connections associated with Electromagnetism [38]. Just as in General Relativity, all physical measurements are relative, so also could the norm of a physical vector depend on its location? If so, a new connection would be required to relate the lengths of the vectors at different positions. This clearly would be a local property. It was called Gauge Invariance. Let us see how this can be expressed mathematically [28, 40]. In essence we have to multiply the norm of a vector $f^\mu(x^\mu) \equiv f(x)$ at $x \equiv x^\mu$ by a scale factor $S(x^\mu) \equiv S(x)$, which latter would represent the change in scale from point to point. So we have for a small displacement to the point $x + dx$, the equations

$$S(x + dx) = 1 + \partial_\mu S dx^\mu$$

$$Sf = f + (\partial_\mu S) f dx^\mu + \partial_\mu f dx^\mu$$

If f is a constant vector, then we have on the right

$$(1 + \partial_\mu S) f dx^\mu$$

As can be seen from the above, the derivative $\partial_\mu S$ is the new mathematical connection associated with the gauge transformation. Weyl identified this connection with the electromagnetic potential A_μ. This is motivated by the fact that a second gauge change with a scale factor Λ leads to

$$\partial_\mu S \to \partial_\mu S + \partial_\mu \Lambda$$

which mimics the behavior under a gauge transformation of the electromagnetic potential in classical theory,

$$A_\mu \to A_\mu + \partial_\mu \Lambda$$

With the advent of Quantum Theory, Weyl himself realized that his old idea could be given a new interpretation. Rather than being a change of scale, a gauge transformation could be interpreted as a phase transformation. This is because if

$$\psi \to \psi e^{-\imath\lambda} \tag{1.12}$$

then for the electromagnetic potential we would have

$$A_\mu \to A_\mu - \partial_\mu \lambda \tag{1.13}$$

Equation (1.12) together with equation (1.13) is a symmetry transformation of the Schrödinger equation. All this is nothing but the well known minimum coupling algorithm,

$$p_\mu \to p_\mu - eA_\mu$$

The reason that this reinterpretation of gauge transformations is acceptable is that the Quantum Mechanical phase is not a directly measurable quantity. It is now clear that Electromagnetism can be interpreted as a Quantum Mechanical local gauge theory. This time it is the local phase of the wave function which is the physical degree of freedom that depends on its spacetime position.

The modern rebirth of gauge theory stemmed from a study of the strong forces mediated by the Yukawa Meson, and Heisenberg's iso spin interpretation of the identity of neutrons and protons when electromagnetic interactions are switched off. That is the strong force was invariant in the SU(2) isotopic spin group.

The difficulty was that iso spin is not a local gauge symmetry, because it is an internal Quantum number independent of spacetime location. So there was no question of an iso spin potential connection whose Quantum would be the Yukawa Meson.

Nevertheless in 1954 Yang and Mills went ahead to treat strong interactions as a gauge invariant field theory by postulating that the local gauge group was the SU(2) iso spin group, in analogy with the electromagnetic case. This time the proposed connection was a linear combination of the angular momentum operators,

$$A_\mu = \sum_\imath A^\imath_\mu(x) L_\imath \tag{1.14}$$

This is a generalization of the electromagnetic case. In the latter, the operators $L_\imath$ are replaced by the unit matrix and the coefficients $A_\mu(x)$ are proportional to the phase change $\delta_\mu \lambda$. As can be seen from (1.14) the Yang-Mills potential is both a field in spacetime and an operator in iso spin space. It must be observed that like the electromagnetic field the Yang-Mills field is mediated by zero mass Bosons. This is because a massive intermediary would imply a term of the form $m^2 A_\mu A^\mu$, which is clearly not gauge invariant.

Let us now see how a symmetry group transformation leads us to a connection which can be identified with the gauge potential field. Indeed, for an arbitrary non-Abelian group, the symmetry transformation is given by

$$U\Psi = \exp\left(\imath q \sum_k \Theta^k(x) F_k\right)\Psi \tag{1.15}$$

In (1.15), the fact that $\Theta^k(x)$ are continuous functions of x defines the local transformation. q is the coupling constant for the gauge group in question. F_k are the generators of the internal symmetry group, satisfying the commutation relations

$$[F_\imath, F_\jmath] = \imath \epsilon_{\imath\jmath k} F_k \, ,$$

In (1.15) if an infinitesimal transformation of the spacetime coordinate is carried out, we get instead of the usual derivative, the gauge covariant derivative describing the changes in both the external and internal components of $\Psi(x)$ viz.,

$$D_\mu \Psi_\beta = \sum_\alpha \left[\delta_{\beta\alpha}\partial_\mu - \imath q (A_\mu)_{\beta\alpha}\right]\Psi_\alpha \tag{1.16}$$

where A_μ are given by

$$(A_\mu)_{\alpha\beta} = \sum_k (\partial_\mu \Theta^k)(F_k)_{\alpha\beta}$$

A special case of (1.16) is the U(1) electromagnetic gauge group, for which this reduces to the usual form with the minimal coupling

$$D_\mu \Psi = (\partial_\mu - \imath q A_\mu)\Psi$$

Thus for the electromagnetic gauge group the gauge covariant derivative is the familiar canonical momentum. It must be noted that the potential A_μ is both an external field and as well, an internal space operator. Furthermore in the non-Abelian gauge group, an internal operator part of the potential would contain a linear combination of the group generators, F_k

which do not in general commute. However as we saw above, the problem has been that we cannot incorporate a mass for the gauge field in an invariant manner. This is achieved by considering an additional field. In the case of weak interaction this is the Higgs field, which breaks the symmetry and leads to a mass generating mechanism.
The Theory of Relativity (Special and General) and Quantum Theory have been often described as the two pillars of twentieth century physics. Each in its own right explained aspects of the universe to a certain extent. But there are still many unanswered questions. For example spacetime singularities (like the Big Bang), termed by John Wheeler as the Greatest Crisis of Physics, the many divergences encountered in particle physics, some eighteen arbitrary parameters in the standard model, elusive monopoles (and Higgs bosons), gravitational waves and Dark Matter and and Supersymmetric particles and so on.
To quote 't Hooft (drawing a comparison with planetary orbits) [41], "What we do know is that the standard model, as it stands today, cannot be entirely correct, in spite of the fact that the interactions stay weak at ultra-short distance scales. Weakness of the interactions at short distances is not enough; we also insist that there be a certain amount of stability. Let us use the metaphor of the planets in their orbits once again. We insisted that, during extremely short time intervals, the effects of the forces acting on the planets have hardly any effect on their velocities, so that they move approximately in straight lines. In our present theories, it is as if at short time intervals several extremely strong forces act on the planets, but, for some reason, they all but balance out. The net force is so weak that only after long time intervals, days, weeks, months, the velocity change of the planets become apparent. In such a situation, however, a reason must be found as to why the forces at short time scales balance out. The way things are for the elementary particles, at present, is that the forces balance out just by accident. It would be an inexplicable accident, and as no other examples of such accidents are known in Nature, at least not of this magnitude, it is reasonable to suspect that the true short distance structure is not exactly as described in the standard model, but that there are more particles and forces involved, whose nature is as yet unclear."
Further, there has been much talk about going beyond the Standard Model, ever since the mass of the neutrino, predicted independently by the author [42] was confirmed by the Super Kamiokande experiment. For according to the Standard Model, the neutrino should be massless. Clearly, we have reached the limits of the Standard Model.

Returning to the issue of Quantum Mechanics and General Relativity it was almost as if Rudyard Kipling's "The twain shall never meet" was true for these two intellectual achievements, a view endorsed by Pauli, who went as far as to say that we should not try to put together what God had intended to be separate. For decades there have been fruitless attempts to unify electromagnetism and gravitation, or Quantum Theory and General Relativity. For, we cannot leave the Universe with a split personality——one for the micro world and one for the macro universe. Such a dichotomic description of nature is totally unsatisfactory. For sometime it looked like String Theory would answer all questions, as we will see shortly.

1.9 Standard Cosmology

In the sixties, it was not suspected that Elementary Particle Physics would be intimately connected with Cosmology, which was at the other end of the spectrum in terms of sizes! But it was subsequently realized that further experimentation on theoretical particle models would require energies that could not be available in foreseeable particle accelerators. Fortunately the Big Bang model of cosmology provides a scenario in the early Universe where such high energies were accessible and consequently particle physics predictions become testable. The very interesting development that has emerged is that Particle Physics and Cosmology have got linked by this high energy bridge.

The so called Big Bang model arose from three main observations. The first was the discovery in the 1920s that the Universe is expanding, in the sense that the basic constituents, the galaxies (as then believed) showed red shifts. Furthermore as Hubble discovered, the farther the galaxy, the greater its speed of recession. This is Hubble's Law: $v = Hr$, where H is the Hubble constant.

Another important observation was about light element abundance——or overabundance——in the Universe. In the 1940s Gamow and coworkers provided an explanation for this. The early Universe must have been very hot and dense. The synthesis of light elements took place when the Universe was at a temperature of $10^9 K$. However heavier elements were formed later, inside the stars, and were strewn about by supernova explosions.

Finally there was a cosmic footprint of an explosion from a very early hot and dense state. This was the residual background radiation from that

early event. In the present epoch however the earlier intense radiation would have cooled, and it was calculated that it would be in the form of microwaves. Exactly such a cosmic background microwave radiation footprint was accidentally discovered in 1965 by Penzias and Wilson. This effectively overthrew a competing model of that time−−the Steady State Model, which has now become history [43].
So the picture to emerge [43–45] was that the Universe was born in a titanic explosion or Big Bang, a name made popular by Gamow. Exactly at the time of the Big Bang some fourteen billion years ago, it is reckoned, all the matter and energy of the Universe was concentrated at a single point, where the density and curvature would be infinite. This is the Big Bang singularity. Following the Big Bang, matter and energy has been flung all round and even today the galaxies (or clusters of galaxies) are rushing outward due to that initial impact.
The question that arises is, will the expansion of the Universe continue for ever, or would it slow down to a halt and then collapse? The answer to this would depend on the mass/energy density of the Universe. If this value is greater than a critical value, then the gravitational attraction will ultimately prevail over the expansion and the Universe would collapse. But if the density is less than the critical value, the Universe would go on expanding for ever. This critical density is given by,

$$\rho_{crit} = \frac{3H^2}{8\pi G} = 2 \times 10^{-29} h^2 g/cm^3$$

Observations seem to indicate that the density of the Universe was close to the critical value. Further an observation of the speeds of rotation of the galaxies indicated that the galaxies themselves contained more matter than met the eye. This lead to "Dark Matter" being invoked. Dark matter has not been directly detected, nor can it be precisely characterized, even though there have been a number of possible candidates. For example invisible Black Holes or even difficult to detect brown dwarf stars. Exotic massive particles have also been proposed as also massive neutrinos or monopoles. With dark matter thrown in, it was believed that the Universe had the critical density to reverse the expansion.
Though the Big Bang model could explain several observations, there were subtler questions which came to haunt. These were: How come the density of the Universe, which could have been anything, is in fact so close to the critical density in a process spread over billions of years? More precisely such a close critical density today would imply that even after about a billionth of a second after the Big Bang the density was equal to the criti-

cal density accurate to some twenty five decimal places. Alternatively this means that the Universe or space is very flat. This need not have been so. And then the Universe appears uniform on large scales. For instance the cosmic microwave background radiation is uniform in temperature to a high degree of accuracy. How can this be so for regions separated by such vast distances. For since the Big Bang for light itself there has not been enough time to connect them. This is called the horizon problem.
Finally how do we account for the small scale inequalities or lumps in the Universe which we see as galaxies?
In 1981 Alan Guth proposed his inflation Theory ([46, 47]). According to this there was a super fast or super rapid expansion in the early stages of the Universe, so that the size of the Universe exploded to several times its original size within a small fraction of a second.
To put it simply this super fast or exponential expansion flattens out the Universe, thus explaining the first problem. The horizon problem is also accounted for: Due to the super fast expansion or inflation, distant regions were much closer together than with an usual expansion. So they would be at the same temperature. Furthermore Quantum fluctuations in the inflation field would cause fluctuations in density, that is they would seed the formation of galaxies. Finally it may be added that given the inflationary scenario, the fact that exotic particles like magnetic monopoles are not detected is also explained. The rapid inflation would have diluted such particles and made them unobservable.
A time line of the Universe would be [48]

$$1 \qquad t = 10^{-43} secs, \quad T = 10^{32} K$$

The Planck era of Quantum Gravity would have just ended and the Universe would be described by a Grand Unified Theory

$$2 \qquad t = 10^{-35} secs, \quad T = 10^{28} K$$

The Grand Unified symmetry is broken. The size of the Universe would still be only a millimeter across

$$3 \qquad t = 10^{-10} secs, \quad T = 10^{15} K$$

At this stage electroweak symmetry is broken. Already the Universe has swelled to a size of $10^{14} cms$.

$$4 \qquad t = 10^{-5} secs, \quad T \sim 10^{12} K$$

QCD is switched off and quarks combine to form hadrons

$$5 \qquad t \sim 3min, \quad T \sim 10^{9} K$$

Nucleosynthesis begins and nuclei of lighter elements like Helium and Lithium begin to form

$$6 \qquad t = 10^{-5} yrs, \quad T \sim 4000K$$

Electrons and nuclei combine to form neutral atoms as charged particles are no longer present. So there is no scattering of photons and radiation in general including the Cosmic Microwave Background Radiation.
Interestingly Optical and Radio Astronomy cannot probe beyond this time

$$7 \qquad t \sim 10^{9} yrs, \quad T \sim 10K$$

Galaxy formation begins

$$8 \qquad t \sim 10^{10} yrs, \quad T \sim 2.7K$$

This is the Universe of today.
The above was the model till 1997. That year, the author put forward an alternative model which in fact went against the then existing belief. On the contrary, this model predicted a dark energy driven accelerating ever expanding Universe. In 1998 dramatic confirmation for the new model came from the observations of Perlmutter, Schmidt, Kirshner and others. We will come back to all this in Chapter 3. Clearly the limits of Standard Big Bang Cosmology had been reached in 1997.

1.10 Bosonic Strings

We have already noted that String Theory (and its derivatives) held the promise of unifying gravitatioin with other fundamental forces. Let us begin with T. Regge's work of the 1950s referred to earlier [32, 49, 50] in which he carried out a complexification of the angular momentum and analyzed particle resonances. As is well known, the resonances could be fitted by a straight line plot in the (J, M^2) plane, where J denotes the angular momentum and M the mass of the resonances. That is we have

$$J \propto M^2, \tag{1.17}$$

Equation (1.17) suggested that not only did resonances have angular momentum, but they also resembled extended objects. This was contrary to the belief that elementary particles were point like. In fact as we saw, at the turn of the twentieth century, Poincare, Lorentz, Abraham and others had toyed with the idea that the electron had a finite extension, but they had to abandon this approach, because of a conflict with Special Relativity.

The problem is that if there is a finite extension for the electron then forces on different parts of the electron would exhibit a time lag, requiring the so called Poincare stresses for stability [6, 7, 33].
In this context, it may be mentioned that in the early 1960s, Dirac came up with an imaginative picture of the electron, not so much as a point particle, but rather a tiny closed membrane or bubble. Further, the higher energy level oscillations of this membrane would represent the "heavier electrons" like muons [51].
Then, in 1968, G. Veneziano came up with a unified description of the Regge resonances (1.17) and other scattering processes. Veneziano considered the collision and scattering process as a black box and pointed out that there were in essence, two scattering channels, s and t channels. These, he argued gave a dual description of the same process [52, 53].
In an s channel, particles A and B collide, form a resonance which quickly disintegrates into particles C and D. On the other hand in a t channel scattering, particles A and B approach each other, and interact via the exchange of a particle q. The result of the interaction is that particles C and D emerge. If we now enclose the resonance and the exchange particle q in an imaginary black box, it will be seen that the s and t channels describe the same input and the same output: They are essentially the same.
There is another interesting hint which we get from Quantum Chromo Dynamics that we encountered. Let us come to the inter-quark potential [30, 34]. There are two interesting features of this potential as noted. The first is that of confinement, which is given by a potential term like

$$V(r) \approx \sigma r, \quad r \to \infty,$$

where σ is a constant. This describes the large distance behavior between two quarks. The confining potential ensures that quarks do not break out of their bound state, which means that effectively free quarks cannot be observed.
The second interesting feature is asymptotic freedom. This is realized by a Coulumbic potential

$$V_c(r) \approx -\frac{\propto (r)}{r} (\text{small}\, r)$$

$$\text{where } \propto (r) \sim \frac{1}{ln(1/\lambda^2 r^2)}$$

The constant σ is called the string tension, because there are string models which yield $V(r)$. This is because, at large distances the inter-quark field is string like with the energy content per unit length becoming constant. Use of the angular momentum−−mass relation indicates that $\sigma \sim (400MeV)^2$. Such considerations lead to strings which are governed by the equation [54–57]

$$\rho\ddot{y} - Ty'' = 0, \tag{1.18}$$

$$\omega = \frac{\pi}{2l}\sqrt{\frac{T}{\rho}}, \tag{1.19}$$

$$T = \frac{mc^2}{l}; \quad \rho = \frac{m}{l}, \tag{1.20}$$

$$\sqrt{T/\rho} = c, \tag{1.21}$$

T being the tension of the string, l its length and ρ the line density and ω in (1.19) the frequency. The identification (1.19),(1.20) gives (1.21), where c is the velocity of light, and (1.18) then goes over to the usual d'Alembertian or massless Klein-Gordon equation.

Further, if the above string is quantized canonically, we get

$$\langle \Delta x^2 \rangle \sim l^2. \tag{1.22}$$

Thus the string can be considered as an infinite collection of harmonic oscillators [55]. (Indeed, we will return to this model of a collection of Harmonic Oscillators, in later Chapters.) Further we can see, using equations (1.19) and (1.20) and the fact that

$$\hbar\omega = mc^2$$

that the extension l is of the order of the Compton wavelength in (1.22), a circumstance that was called one of the miracles of the string theory by Veneziano [52].

It must be mentioned that the above considerations describe a "Bosonic String", in the sense that there is no room for the Quantum Mechanical spin. This can be achieved by giving a rotation to the relativistic quantized string as was done by Ramond [16, 58]. In this case we recover (1.17) of the Regge trajectories. The particle is now an extended object, at the Compton scale, rotating with the velocity of light. Furthermore in superstring theory there is an additional term a_0, viz.,

$$J \leq (2\pi T)^{-1}M^2 + a_0\hbar, \text{ with } a_0 = +1(+2) \text{ for the open (closed) string.} \tag{1.23}$$

The term a_0 in (1.23) comes from the Zero Point Energy. Usual gauge bosons are described by $a_0 = 1$ and gravitons by $a_0 = 2$.
It is also well known that string theory has always had to deal with extra dimensions which reduce to the usual four dimensions of physical space-time when we invoke the Kaluza-Klein approach at the Planck Scale [59]. Briefly, this means that the extra dimensions have infinitesimal dimensions, or more accurately, are wound up in infinitesimal cylinders.
All these considerations have been leading to more and more complex models, the latest version being the so called M-Theory. In this latest theory supersymmetry is broken so that the supersymmetric partner particles do not have the same mass as the known particles. Particles can now be described as soliton like branes, resembling the earlier Dirac membrane. M-Theory also gives an interface with Black Hole Physics. Further these new masses must be much too heavy to be detected by current accelerators. The advantage of Supersymmetry (SUSY) is that a framework is now available for the unification of all the interactions including Gravitation. It may be mentioned that under a SUSY transformation, the laws of physics are the same for all observers, which is the case in General Relativity (Gravitation) also. Under SUSY there can be a maximum of eleven dimensions, the extra dimensions being curled up as in Kaluza-Klein theories. In this case there can only be an integral number of waves around the circle, giving rise to particles with quantized energy. However for observers in the other four dimensions, it would be quantized charges, not energies. The unit of charge would depend on the radius of the circle, the Planck radius yielding the value e. This is the root of the unification of electromagnetism and gravitation in these theories.
In M-Theory, the position coordinates become matrices and this leads to, as we will see, a noncommutative geometry or fuzzy spacetime in which spacetime points are no longer well defined [60]

$$[x, y] \neq 0$$

From this point of view the mysterious M in M-Theory could stand for Matrix, rather than Membrane or magic as some suppose.
So M-Theory is the new avatar of Quantum Superstring Theory. Nevertheless it is anything but the last word. There are still any number of routes for compressing ten dimensions to our four dimensions. There is still no contact with experiment. It also appears that these theories lead to an unacceptably high cosmological constant and so on and so on. This has prompted some String theorists to invoke an Anthropic Principle approach.

According to this, the universe is really a landscape of some 10^{500} universes, each with its own fundamental laws and constants. It so happens that we are inhabiting a universe with the observed laws and physical constants. This could well be the death knell of the theory.
All this along with the non-verifiability of the above considerations and the fact that the Planck scale $\sim 10^{20} GeV$ is also beyond forseeable attainment in collidors has lead to much criticism even though it is generally accepted that the mathematical ideas have been rich and promising. Indeed it is becoming increasingly clear that this intellectual tour de force, touted as The Theory of Everything has perhaps reached its limits. Not just String Theory, but also all so called "reductionist" theories are under the scanner.

1.11 End of the Road?

Reductionism has been at the heart of twentieth century Theoretical Physics. Beginning with the atomism of ancient India and Greek thinkers, it was reborn in the nineteenth century. This spirit is very much evident in Einstein's concept of locality in which an arbitrarily small part of the universe can be studied without reference to other parts of it. Indeed it is this philosophy of reductionism which has propelled the most recent studies such as String Theory or other Quantum Gravity approaches.
Against this backdrop, the first salvo was fired by Nobel Laureate R.B.Laughlin. "A Different Universe", his recent book, would come as a shock because he debunks reductionism in favour of what is these days called emergence. That is his central theme [61]. The fundamental laws of nature emerge through collective self organization and do not require knowledge of their component parts, that is microscopic rules, in order that we comprehend or exploit them. The distinction between fundamental laws and laws descending from them is a myth. In his words, "... I must openly discuss some shocking ideas: the vacuum of space-time is 'matter', the possibility that relativity is not fundamental..." He argues that all fundamental constants require an environmental context to make sense. This is contrary to the reductionist view that basic bricks build up structures.
Laughlin takes pain to bring to our notice that there is now a paradigm shift from the older reductionist view to a view of emergence. For him Klitzing's beautiful experiment bringing out the Quantum Hall effect is symbolic of the new ethos. The Quantum of Hall resistance is a combination of fundamental constants viz., the indivisible quantum of electric

charge e, the Planck constant h and the speed of light c. This means that these supposedly basic building blocks of the universe can be measured with breathtaking accuracy, without dealing with the building blocks themselves. Though, from one point of view, this resembles the fact that bulk properties emerge from underlying and more fundamental microscopic properties, he argues that this latter effect reveals that supposedly indivisible quanta like the electric charge e can be broken into pieces through self organization of phases. That is, the supposedly fundamental things are not necessarily fundamental. Furthermore, for example, in superconductivity, many of the so called minor details are actually inessential−−the exactness of the Meissner and Josephson effects does not require the rest of the finer detail to be true.

Admittedly there are a number of grey areas in modern theoretical physics which are generally glossed over. For instance Dirac's Hole Theory of anti matter. However in silicon, there are many electrons locked up in the chemical bonds and it is possible to pull an electron out of a chemical bond. This makes a hole which is mobile and acts in every way like an extra electron with opposite charge added to the silicon. This idea however requires the analogue of a solid's bond length. In Particle Physics such a length conflicts fundamentally with the Principle of Relativity as we have seen, unless, as we have argued, it breaks down at the Compton scale. On the contrary, Laughlin laments, "... instead, physicists have developed clever semantic techniques for papering it over... Thus instead of Holes one speaks of anti particles. Instead of bond length one speaks of an abstraction called the ultra violet cutoff, a tiny length scale introduced into the problem to regulate−−which is to say, to cause it to make sense. Below this scale one simply aborts one's calculations... Much of Quantum Electrodynamics, the mathematical description of how light communicates with the ocean of electrons... boils down to demonstrating the unmeasurableness of the ultra violet cutoff... The potential of overcoming the ultra violet problem is also the deeper reason for the allure of String Theory, a microscopic model for the vacuum that has failed to account for any measured thing... The properties of empty space relevant to our lives show all the signs of being emergent phenomena characteristic of a phase of matter. They are simple, exact, model insensitive, and universal. This is what insensitivity to the ultra violet cutoff means physically."

Moreover as we will see particularly in Chapter 4, quantized sound waves or phonons have an exact parallel with photons−−in fact their quantum properties are identical to those of light. However sound is a collective motion of elastic matter, while in our understanding, light is not. This means that quantization of sound may be deduced from the underlying laws of Quantum Mechanics obeyed by the atoms, whereas in the case of light this is postulated. This is a logical loose end and ultimately we bring in the gauge effect to cover this. But unfortunately, "there is also a fundamental incompatibility of the gauge effect with the principle of relativity, which one must sweep under the rug by manipulating the cutoff." Laughlin complains that in spite of the evidence against reductionism, sub nuclear experiments are generally described in reductionist terms. These points will be considered in detail, in the following Chapters.

Turning to General Relativity, Laughlin points out that it is a speculative post Newtonian Theory of Gravity, an invention of the mind, "it is just controversial and largely beyond the reach of experiment", unlike Special Relativity which was a discovery of the behavior of nature. He then points out the contradiction between Special and General Relativity−−in the former Einstein did away with the concept of the Ether. But this reenters the latter theory in the form of the fabric of space. Touching upon the skeletons in the closet of General Relativity, Laughlin discusses the embarrassment that is caused by a non zero cosmological constant.

He concludes that if Einstein were alive today, he would be horrified at this state of affairs and would conclude that his beloved principle of Relativity was not fundamental at all but emergent.

Laughlin takes a critical look at renormalizability, a pillar of modern theoretical physics, and cosmology. Indeed, we have already commented on this. "If renormalizability of the vacuum is caused by proximity to phase transitions, then the search for an ultimate theory would be doomed on two counts: It would not predict anything even if you found it, and it could not be falsified...

"The political nature of cosmological theories explains how they could so easily amalgamate String Theory, a body of mathematics with which they actually have very little in common... (String Theory) has no practical utility however, other than to sustain the myth of the ultimate theory. There is no experimental evidence for the existence of Strings in nature... String Theory is, in fact a textbook case of ... a beautiful set of ideas that will always remain just barely out of reach. Far from a wonderful technological hope for a greater tomorrow, it is instead a tragic consequence of an absolute belief system in which emergence plays no role..."

Laughlin has captured the mood of pessimism that prevails in the minds of several high energy physicists. He goes on to cite the famous joke that the

Physical Review is now so voluminous that stacking up successive issues would generate a surface travelling faster than the speed of light, although without violating Relativity, because the Physical Review contains no information anyway.

However before proceeding further it must be mentioned, as we will see in the following Chapters, that the author's own work during the last decade has borne out the spirit of these ideas, that the iron clad law of physics is more thermodynamic and stochastic in nature, that the velocity of light or the gravitational constant can be deduced from such considerations rather than be taken as fundamental inputs; how, it is possible to have schemes that bypass the awkward questions raised in the book, without brushing them away below the carpet, by considering an a priori Quantum vacuum in which fluctuations take place.

Returning now to String Theory there is no doubt that it has straddled the past two decades and more as the only contender for the Theory of Everything. Some years ago Nobel Laureate Sheldon Glashow described it, sarcastically, as the only game in town. In recent months though the theory is not only being debunked, it is facing a lot of flak, particularly in the worldwide media. Laughlin may be faulted on the grounds that he is not a String Theorist or a Particle Physicist. The decisive tilt has come from Nobel Laureate David Gross, very much an insider, who as it were, spilt the beans at the recent 23rd Solvay Conference in Physics held in Brussels, Belgium, in late 2005. He stated "We don't know what we are talking about!" He then went on to say, "Many of us believed that string theory was a very dramatic break with out previous notions of quantum theory. But now we learn that string theory, well, is not that much of a break." And that physics is in "a period of utter confusion."

At this meeting Gross compared the state of Physics today to that during the first Solvay Conference in 1911 "They were missing something absolutely fundamental," he said. "We are missing perhaps something as profound as they were back then."

Thus the Time Magazine, August 14, 2006 issue notes:

"By now, just about everyone has heard of string theory. Even those who don't really understand it-which is to say, just about everyone-know that it's the hottest thing in theoretical physics. Any university that doesn't have at least one string theorist on the payroll is considered a scientific backwater. The public, meanwhile, has been regaled for years with magazine articles

"But despite its extraordinary popularity among some of the smartest people on the planet, string theory hasn't been embraced by everyone-and now, nearly 30 years after it made its initial splash, some of the doubters are becoming more vocal. Skeptical bloggers have become increasingly critical of the theory, and next month two books will be hitting the shelves to make the point in greater detail. Not Even Wrong, by Columbia University mathematician Peter Woit, and The Trouble With Physics, by Lee Smolin at the Perimeter Institute for Theoretical Physics in Waterloo, Ont., both argue that string theory (or superstring theory, as it is also known) is largely a fad propped up by practitioners who tend to be arrogantly dismissive of anyone who dare suggest that the emperor has no clothes

"Bizarre as it seemed, this scheme appeared on first blush to explain why particles have the characteristics they do. As a side benefit, it also included a quantum version of gravity and thus of relativity. Just as important, nobody had a better idea. So lots of physicists, including Woit and Smolin, began working on it.

"Since then, however, superstrings have proved a lot more complex than anyone expected. The mathematics is excruciatingly tough, and when problems arise, the solutions often introduce yet another layer of complexity. Indeed, one of the theory's proponents calls the latest of many string-theory refinements "a Rube Goldberg contraption." Complexity isn't necessarily the kiss of death in physics, but in this case the new, improved theory posits a nearly infinite number of different possible universes, with no way of showing that ours is more likely than any of the others.

"That lack of specificity hasn't slowed down the string folks. Maybe, they've argued, there really are an infinite number of universes-an idea that's currently in vogue among some astronomers as well-and some version of the theory describes each of them. That means any prediction, however outlandish, has a chance of being valid for at least one universe, and no prediction, however sensible, might be valid for all of them.

"That sort of reasoning drives critics up the wall. It was bad enough, they say, when string theorists treated nonbelievers as though they were a little slow-witted. Now, it seems, at least some superstring advocates are ready to abandon the essential definition of science itself on the basis that string theory is too important to be hampered by old-fashioned notions of experimental proof

"And it is that absence of proof that is perhaps most damning."

"It's fine to propose speculative ideas," says Woit, "but if they can't be tested they're not science." To borrow the withering dismissal coined by the great physicist Wolfgang Pauli, they don't even rise to the level of being wrong. That, says Sean Carroll of the University of Chicago, who has worked on strings, is unfortunate. "I wish string theorists would take the goal of connecting to experiment more seriously," he says.

According to the August 27, 2006 issue of the Scientific American,

"With a tweak to the algorithms and a different database, the Website could probably be made to spit out what appear to be abstracts about superstring theory: "Frobenius transformation, mirror map and instanton numbers" or "Fractional two-branes, toric orbifolds and the quantum McKay correspondence." Those are actually titles of papers recently posted to arXiv.org repository of preprints in theoretical physics, and they may well be of scientific worth-if, that is, superstring theory really is a science. Two new books suggest otherwise: that the frenzy of research into strings and branes and curled-up dimensions is a case of surface without depth, a solipsistic shuffling of symbols as relevant to understanding the universe as randomly generated dadaist prose.

"In this grim assessment, string theory-an attempt to weave together general relativity and quantum mechanics−−is not just untested but untestable, incapable of ever making predictions that can be experimentally checked. With no means to verify its truth, superstring theory, in the words of Burton Richter, director emeritus of the Stanford Linear Accelerator Center, may turn out to be "a kind of metaphysical wonderland." Yet it is being pursued as vigorously as ever, its critics complain, treated as the only game in town.

"String theory now has such a dominant position in the academy that it is practically career suicide for young theoretical physicists not to join the field," writes Lee Smolin, a physicist at the Perimeter Institute for Theoretical Physics, in The Trouble with Physics: The Rise of String Theory, the Fall of a Science, and What Comes Next. "Some young string theorists have told me that they feel constrained to work on string theory whether or not they believe in it, because it is perceived as the ticket to a professorship at a university."

"Neither of these books can be dismissed as a diatribe. Both Smolin and Woit acknowledge that some important mathematics has come from contemplating superstrings. But with no proper theory in sight, they assert, it is time to move on. "The one thing everyone who cares about fundamental physics seems to agree on is that new ideas are needed," Smolin writes. "We are missing something big."

"The story of how a backwater of theoretical physics became not just the rage but the establishment has all the booms and busts of an Old West min-ing town. Unable to fit the four forces of nature under the same roof, a few theorists in the 1970s began adding extra rooms––the seven dimensions of additional closet space that unification seemed to demand. With some mathematical sleight of hand, these unseen dimensions could be curled up ("compactified") and hidden inside the cracks of the theory, but there were an infinite number of ways to do this. One of the arrangements might describe this universe, but which?

"The despair turned to excitement when the possibilities were reduced to five and to exhilaration when, in the mid-1990s, the five were funneled into something called M Theory, which promised to be the one true way. There were even hopes of experimental verification

"That was six years ago, and to hear Smolin and Woit tell it, the field is back to square one: recent research suggests that there are, in fact, some 10^{500} perfectly good M theories, each describing a different physics. The theory of everything, as Smolin puts it, has become a theory of anything.

"Faced with this free-for-all, some string theorists have concluded that there is no unique theory, that the universe is not elegant but accidental. If so, trying to explain the value of the cosmological constant would make as much sense as seeking a deep mathematical reason for why stop signs are octagonal or why there are 33 human vertebrae"

An article in the Financial Times (London) in June 2006 by Physicist Robert Mathews noted:

"They call their leader The Pope, insist theirs is the only path to enlightenment and attract a steady stream of young acolytes to their cause. A crackpot religious cult? No, something far scarier: a scientific community that has completely lost touch with reality and is robbing us of some of our most brilliant minds.

"Yet if you listened to its cheerleaders-or read one of their best-selling books or watched their television mini-series-you, too, might fall under their spell. You, too, might come to believe they really are close to revealing the ultimate universal truths, in the form of a set of equations describing the cosmos and everything in it. Or, as they modestly put it, a "theory of everything".

"This is not a truth universally acknowledged. For years there has been concern within the rest of the scientific community that the quest for the theory of everything is an exercise in self-delusion. This is based on the simple fact that, in spite of decades of effort, the quest has failed to produce a single testable prediction, let alone one that has been confirmed.

"For many scientists, that makes the whole enterprise worse than a theory that proves to be wrong. It puts it in the worst category of a scientific theories, identified by the Nobel Prize-winning physicist Wolfgang Pauli: it is not even wrong. By failing to make any predictions, it is impossible to tell if it is a turkey, let alone a triumph.

"It is this loss of contact with reality that has prompted so much concern among scientists-at least, those who are not intimidated by all the talk of multidimensional superstrings and Calabi-Yau manifolds that goes with the territory. But now one of them has decided the outside world should be told about this scientific charade. As a mathematician at Columbia University, Peter Woit has followed the quest for the theory of everything for more than 20 years. In his new book "Not Even Wrong" he charts how a once-promising approach to the deepest mysteries in science has mutated into something worryingly close to a religious cult."

A review in the December 2006 issue of Physics Today notes: "Noted theoretical physicist Sheldon Glashow has famously likened string theory to medieval theology because he believes both are speculations that cannot be tested. Yet if readers believe Lee Smolin and Peter Woit, they might conclude that the more apt comparison is to the Great disappointment of 1844, when followers of the Baptist preacher William Miller gave up all their worldly possessions and waited for the Second Coming. The empirical inadequacy of that prediction led to apostasy and schisms among thousands of Miller's followers. At least one of the branches claimed that the event had in fact occurred, but in a heavenly landscape linked to the world of experience through only the weak but all-pervasive spiritual interaction. Yet irritating differences exist between Miller's followers and the "disappointed" of the 1984 coming of the theory of everything; a majority of the latter seem to have preserved their faith and gained worldly fortune in the form of funding, jobs, and luxurious conferences at exotic locales."

In all this confusion, we should not forget two important points. The first is that String Theory still remains a mathematically beautiful self consistent system of thought. Perhaps the flak that String Theory is receiving is more due to reasons in the domain of the sociology of science. To elaborate, String Theory has been touted as a theory, which, in the strict sense it is not. If it had been promoted as a hypothesis, one of a few possible, perhaps, there would have been much less criticism.

Indeed, some years ago Nobel Laureate 't Hooft had noted [62] "Actually, I would not even be prepared to call string theory a "theory" but rather a "model" or not even that: just a hunch. After all, a theory should come together with instructions on how to deal with it to identify the things one wishes to describe, in our case the elementary particles, and one should, at least in principle, be able to formulate the rules for calculating the properties of these particles, and how to make new predictions for them."

Moreover in the process String Theory adopted strong arm facist type tactics including marketing through the media, while at the same time making not too covert attempts to suppress other ideas. The backlash was therefore inevitable.

The second point is even more important and is expressed in David Gross's statement that perhaps we are missing something very profound. This throws up a great challenge and makes for very exciting times.

Science has been described as a quest for the how and why of nature. Over the centuries it has been guided by some principles which have crystallized into a methodology. Thus observation leads to the framing of hypotheses. It is expected that the hypotheses would have maximum simplicity and maximal economy. This means that, apart from being simple, the maximum number of observations are explained by a minimum number of hypotheses. Further tests would then confirm or disprove the hypotheses, if the hypotheses are found to be consistent with experiment. The richness of a hypothesis is judged by the predictions it can make. These predictions must be either provable or disprovable as stressed by Sir Karl Popper.

As far as Fundamental Physics is concerned, starting from the early days of Indian and later Greek Atomism, through the Atomic Theory of the 19th century, and subsequently the developments in the 20th century, and the early part of the 21st century, the route followed has been one of a descending-in-size cascade. We have been propelled by the belief that the universe could be understood by a study of its ultimate subconstituents. This spirit as noted is very much evident in Einstein's concepts of locality in which an arbitrarily small part of the universe can be studied without reference to other parts of it. A few decades later Wheeler observed that our studies of the inaccessible Planck scale of $10^{-33} cms$ were really like an understanding of bulk properties of matter by studying the subconstituent molecules [45].

Indeed it is this philosophy of reductionism which has propelled the most recent studies such as String Theory or other Quantum Gravity approaches. Decades of labour has gone into these endeavours and the research output has been enormous.

Nevertheless we seem to have reached an impasse of a type that is all too familiar from the past. There are minor discrepancies or corrections, which nevertheless would point to, not just an incremental change of our concepts, but rather to a paradigm shift itself of the type we witnessed with Kepler's ellipses or Einstein's Relativity or Quantum Theory. Any theory can go only so far as its inherent limitations or constraints permit. At that stage, as Thomas Kuhn [63] notes, there would be a revolution, an overturn of concepts and the old way of looking at things. It is no longer an incremental improvement. It is proposed in this book that the so far one way street of reductionism has reached such a limit, and now has to be tempered in the spirit of Thermodynamics or emergence.

Chapter 2

Law Without Law

2.1 A "Lawless" Universe?

From the beginning of modern science, the universe has been considered to be governed by rigid laws which therefore, in a sense, made the universe somehow deterministic. However, it would be more natural to expect that the underpinning for these laws would be random,unpredictable and spontaneous rather than enforced events. This alternative but historical school of thought is in the spirit of Prigogine's, "Order out of chaos"[4].
Prigogine notes, "As we have already stated, we subscribe to the view that classical science has now reached its limit. One aspect of this transformation is the discovery of the limitations of classical concepts that imply that a knowledge of the world "as it is" was possible. The omniscient beings, Laplace's or Maxwell's demon, or Einstein's God, beings that play such an important role in scientific reasoning, embody the kinds of extrapolation physicists thought they were allowed to make. As randomness, complexity, and irreversibility enter into physics as objects of positive knowledge, we are moving away from this rather naive assumption of a direct connection between our description of the world and the world itself. Objectivity in theoretical physics takes on a more subtle meaning. ...Still there is only one type of change surviving in dynamics, one "process", and that is motion... It is interesting to compare dynamic change with the atomists' conception of change, which enjoyed considerable favor at the time Newton formulated his laws. Actually, it seems that not only Descartes, Gessendi, and d'Alembert, but even Newton himself believed that collisions between hard atoms were the ultimate, and perhaps the only, sources of changes of motion. Nevertheless, the dynamic and the atomic descriptions differ radically. Indeed, the continuous nature of the acceleration described by the dynamic equa-

tions is in sharp contrast with the discontinuous, instantaneous collisions between hard particles. Newton had already noticed that, in contradiction to dynamics, an irreversible loss of motion is involved in each hard collision. The only reversible collision–that is, the only one in agreement with the laws of dynamics–is the "elastic," momentum-conserving collision. But how can the complex property of "elasticity" be applied to atoms that are supposed to be the fundamental elements of nature?

"On the other hand, at a less technical level, the laws of dynamic motion seem to contradict the randomness generally attributed to collisions between atoms. The ancient philosophers had already pointed out that any natural process can be interpreted in many different ways in terms of the motion of and collisions between atoms."

In the words of Wheeler[64], we seek ultimately a "Law without Law." Laws are an apriori blue print within the constraints of which, the universe evolves. The point can be understood in the words of Prigogine [65]

"...This problem is a continuation of the famous controversy between Parmenides and Heraclitus. Parmenides insisted that there is nothing new, that everything was there and will be ever there. This statement is paradoxical because the situation changed before and after he wrote his famous poem. On the other hand, Heraclitus insisted on change. In a sense, after Newton's dynamics, it seemed that Parmenides was right, because Newton's theory is a deterministic theory and time is reversible. Therefore nothing new can appear. On the other hand, philosophers were divided. Many great philosophers shared the views of Parmenides. But since the nineteenth century, since Hegel, Bergson, Heidegger, philosophy took a different point of view. Time is our existential dimension. As you know, we have inherited from the nineteenth century two different world views. The world view of dynamics, mechanics and the world view of thermodynamics."

It may be mentioned that subsequent developments in Quantum Theory, including Quantum Field Theory are in the spirit of the former. Einstein himself believed in this view of what may be called deterministic time−−time that is also reversible. On the other hand Heraclitus's point of view was in the latter spirit. His famous dictum was, "You never step into the same river twice", a point of view which was endorsed by earlier ancient Indian thought. This has been the age old tussle between "being" and "becoming".

As Wheeler put it, (loc.cit), "All of physics in my view, will be seen someday to follow the pattern of thermodynamics and statistical mechanics, of regularity based on chaos, of "law without law". Specifically, I believe that everything is built higgledy-piggledy on the unpredictable outcomes of billions upon billions of elementary quantum phenomena, and that the laws and initial conditions of physics arise out of this chaos by the action of a regulating principle, the discovery and proper formulation of which is the number one task...."

The reason this approach is more natural is, that otherwise we would be lead to ask, "from where have these laws come?" unless we either postulate a priori laws or we take shelter behind an anthropic argument. An interesting but neglected body of work in the past few decades is that of Random or Stochastic Mechanics and Electrodynamics. It may be mentioned that a considerable amount of work has been done in this direction by Nelson, Landau, Prugovecki, the author and others[66]-[92], who have tried to derive the Schrodinger equation, the Klein-Gordon equation and even the Dirac equation from stochastic considerations, and in general develop an underpinning of stochastic mechanics and stochastic electrodynamics. The literature is vast and some of the references given cite an extensive bibliography. A few of these approaches have been very briefly touched upon in Cf.ref.[16]. However, all these derivations contain certain assumptions whose meaning has been unclear. We will see examples of this in the sequel. In any case, we will argue that the seeds of a new world view, of the paradigm shift are to be found here in these considerations.

In the above context, we propose below that purely stochastic processes lead to minimum space-time intervals of the order of the Compton wavelength and time, whose considerable significance will be seen and it is this circumstance that underlies quantum phenomena and cosmology, and, in the thermodynamic limit in which N, the number of particles in the universe $\to \infty$, classical phenomena and Quantum Theory as well. In the process, we will obtain a rationale for some of the ad hoc assumptions referred to above.

In the older, and more popular world view, spacetime has generally been taken to be a differentiable manifold with an Euclidean (Galilean) or Minkowskian or Riemannian character. Though the Heisenberg Principle in Quantum Theory forbids arbitrarily small space time intervals, the above continuum character with space time points has been taken for granted even in Quantum Field Theory. In fact if we accept the proposition that what we know of the universe is a result of our measurement (which includes our perception), and that measurements are based on quantifiable units, then it becomes apparent that a continuum is at best an idealization. This was the reason behind the paradox of the point electron which was encountered in the classical theory of the electron, as we saw in the last Chapter. It was also encountered as is well known in Dirac's Quantum Mechanical treatment of the relativistic, spinning electron in which the electron showed up with the velocity of light.

Quantum Mechanics has lived with this self contradiction[93]. In this schizophrenic existence, the wave function follows a deterministic (time reversible) equation, while the result of a measurement, without which no information is retrievable, follows from an acausal "collapse of the wave function" yielding one of the many permissible eigen values, in an unpredictable but probabilistic manner. Indeed it has been suggested by Snyder, Lee and others that the infinities which plague Quantum Field Theory are symptomatic of the fact that space time has a granular or discrete rather than continuous character. This has lead to a consideration of extended particles[94]-[100] [6], as against point particles of conventional theory. Wheeler's space time foam and strings[101]-[103] are in this class, with a minimum cut off at the Planck scale. As 't Hooft notes, [104] "It is somewhat puzzling to the present author why the lattice structure of space and time had escaped attention from other investigators up till now..." We will return to this point later.

All this has also lead to a review of the conventional concept of a rigid background spacetime. More recently [105]-[107], it has been pointed out by the author that it is possible to give a stochastic underpinning to space time and physical laws. This is in the spirit of Wheeler's, "Law without Law" [64] alluded to. In fact in a private communication to the author, Prof. Prigogine wrote, "...I agree with you that spacetime has a stochastic underpinning".

2.2 The Emergence of Spacetime

We will later briefly survey some models for spacetime. For the moment our starting point is the well known fact that in a random walk, the average distance l covered at a stretch is given by [108]

$$l = R/\sqrt{N} \tag{2.1}$$

where R is the dimension of the system and N is the total number of steps. We get the same relation in Wheeler's famous travelling salesman problem and similar problems [109]. The interesting fact that equation (2.1) is true in the universe itself with R the radius of the universe $\sim 10^{28} cm, N$ the number of the elementary particles in the universe $\sim 10^{80}$ and l the Compton wavelength of the typical elementary particle, for example the pion $\sim 10^{-13} cm$ had been noticed a long time ago[110]. From a different point of view, it is one of the cosmic "coincidences" or Large Number relations, pointed out by Weyl, Eddington and others. In this context, equation (2.1) which has been generally considered to be accidental (along with other such relations which we will encounter), will be shown to arise quite naturally in a cosmological scheme based on fluctuations. We would like to stress that we encounter the Compton wavelength as an important and fundamental minimum unit of length and will return recurrently to this theme.

It may be mentioned that a minimum time interval, the chronon, has been considered earlier in a different context by several authors as we will see very soon. What distinguishes Quantum Theory from Classical Physics is as pointed out, the role of the resolution of the observer or observing apparatus. What appears smooth at one level of perception, may turn out to be very irregular on a closer examination. Indeed as noted by Abbot and Wise[111], in this respect the situation is similar to everywhere continuous but non differentiable curves, the fractals of Mandelbrot [112]. This again is tied up with the Random Walk or Brownian character of the Quantum path as noted by Sornette and others[113]-[120]: At scales larger than the Compton wavelength but smaller than the de Broglie wavelength, the Quantum paths have the fractal dimension 2 of Brownian paths (cf. also Nottale,[121]). This will be touched upon briefly in Section 2.6.

This irregular nature of the Quantum Mechanical path was noticed by Feynman [122] "...these irregularities are such that the "average" square velocity does not exist, where we have used the classical analogue in referring to an "average".

"If some average velocity is defined for a short time interval Δt, as, for example, $|x(t+\Delta t) - x(t)|/\Delta t$, the "mean" square value of this is $-\hbar/(m\Delta t)$. That is, the "mean" square value of a velocity averaged over a short time interval is finite, but its value becomes larger as the interval becomes shorter. It appears that quantum-mechanical paths are very irregular. However, these irregularities average out over a reasonable length of time to produce a reasonable drift, or "average" velocity, although for short intervals of time the "average" value of the velocity is very high..."

This as we will see was Dirac's conclusion too, and indeed his explanation for the luminal velocity of the point electron and the non Hermiticity of its position operator in his relativistic electron theory.

Two important characteristics of the Compton wavelength have to be re-emphasized (Cf.[107]): On the one hand with a minimum space time cut off at the Compton wavelength, as we will see later in Chapter 6, we can recover by a simple coordinate shift the Dirac structure for the equation of the electron, including the spin half. In this sense the spin half, which is purely Quantum Mechanical will be seen to be symptomatic of the minimum space time cut off, as is also suggested by the zitterbewegung interpretation of Dirac (in terms of the Uncertainty Principle), Hestenes and others (Cf. discussion in [16]). The zitterbewegung is symptomatic of the fact that by the Heisenberg Uncertainty Principle, physics begins only after an averaging over the minimum spacetime intervals. This is also suggested by stochastic models of Quantum Mechanics referred to, both non relativistic and relativistic as also Feynman's Path Integral formulation. We will comment upon this in the sequel.

On the other hand, we will see that (2.1) and a similar equation for the Compton time in terms of the age of the universe, viz.,

$$T \approx \sqrt{N}\tau \qquad (2.2)$$

can be the starting point for a unified scheme for physical interactions and indeed a cosmology that is not only consistent with observation in which we will deduce the Large Number coincidences referred to, but also predicted in 1997 an accelerating expanding universe when the ruling paradigm was exactly the opposite. We will see this in the next Chapter in detail. The Large Number relations also include a mysterious formula [17], connecting the pion mass and the Hubble constant which we will deduce. It has to be pointed out [109] that in the spirit of Wheeler's travelling salesman's "practical man's minimum" length that the Compton scale plays such a role, and that spacetime is like Richardson's delineation of a jagged coastline [112] with a thick brush, the thickness of the brush being comparable to the Compton scale.

What Richardson found was that the length of the common land boundaries claimed by Portugal and Spain as also Netherlands and Belgium, differed by as much as 20%! The answer to this non-existent border dispute lies in the fact that we are carrying over our concepts of smooth curves or rectifiable arcs to the measurement of real life jagged boundaries or coastlines. As far as these latter are concerned, as Mandelbrot puts it [112] "The result is most peculiar; coastline length turns out to be an elusive notion that slips between the fingers of one who wants to grasp it. All measurement methods ultimately lead to the conclusion that the typical coastline's length is very large and so ill determined that it is best considered infinite....." This is where Hansdorf dimension or the fractal dimension referred to earlier comes in– we are approximating a higher dimensional curve by a one dimensional curve.

Spacetime, rather than being a smooth continuum, is more like a fractal Brownian curve, what may be called Quantized Fractal Spacetime. All this has been recognized by some scholars, at least in spirit. As V.L. Ginzburg puts it [123] "The Special and General Relativity theory, non-relativistic Quantum Mechanics and present theory of Quantum Fields use the concept of continuous, essentially classical, space and time (a point of spacetime is described by four coordinates $x_l = x, y, z, ct$ which may vary continuosly). But is this concept valid always? How can we be sure that on a "small scale" time and space do not become quite different, somehow fragmentized, discrete, quantized? This is by no means a novel question, the first to ask it was, apparently Riemann back in 1854 and it has repeatedly been discussed since that time. For instance, Einstein said in his well known lecture "Geometry and Experience" in 1921: 'It is true that this proposed physical interpretation of geometry breaks down when applied immediately to spaces of submolecular order of magnitude. But nevertheless, even in questions as to the constitution of elementary particles, it retains part of its significance. For even when it is a question of describing the electrical elementary particles constituting matter, the attempt may still be made to ascribe physical meaning to those field concepts which have been physically defined for the purpose of describing the geometrical behavior of bodies which are large as compared with the molecule. Success alone can decide as to the justification of such an attempt, which postulates physical reality for the fundamental principles of Riemann's geometry outside of the domain of their physical definitions. It might possibly turn out that this extrapolation has no better warrant than the extrapolation of the concept of temperature to parts of a body of molecular order of magnitude'.

"This lucidly formulated question about the limits of applicability of the Riemannian geometry (that is, in fact macroscopic, or classical, geometric concepts) has not yet been answered. As we move to the field of increasingly high energies and, hence to "closer" collisions between various particles the scale of unexplored space regions becomes smaller. Now we may possibly state that the usual space relationships down to the distance of the order of $10^{-15}cm$ are valid, or more exactly, that their application does not lead to inconsistencies. It cannot be ruled out that, the limit is nonexistent but it is much more likely that there exists a fundamental (elementary) length $l_0 \leq 10^{-16} - 10^{-17}cm$ which restricts the possibilities of

classical, spatial description. Moreover, it seems reasonable to assume that the fundamental length l_0 is, at least, not less than the gravitational length $l_g = \sqrt{Gh/c^3} \sim 10^{-33} cm$.
"... It is probable that the fundamental length would be a "cut-off" factor which is essential to the current quantum theory: a theory using a fundamental length automatically excludes divergent results".
Einstein himself was aware of this possibility. As he observed [124], "... It has been pointed out that the introduction of a spacetime continuum may be considered as contrary to nature in view of the molecular structure of everything which happens on a small scale. It is maintained that perhaps the success of the Heisenberg method points to a purely algebraic method of description of nature that is to the elimination of continuous functions from physics. Then however, we must also give up, by principle the spacetime continuum. It is not unimaginable that human ingenuity will some day find methods which will make it possible to proceed along such a path. At present however, such a program looks like an attempt to breathe in empty space".
To analyse this further, we observe that space time given by R and T of (2.1) and (2.2) represents a measure of dispersion in a normal distribution: Indeed if we have a large collection of N events (or steps) of length l or τ, forming a normal distribution, then the dispersion σ is given by precisely the relation (2.1) or (2.2).
The significance of this is brought out by the fact that the universe is a collection of N elementary particles, in fact typically pions of "size" l, as seen above. We consider spacetime not as an apriori container of these particles but rather as a Gaussian collection of these particles, a Random Heap. At this stage, we do not even need the concept of a continuum.
In this scheme the probability distribution has a width or dispersion $\sim \frac{1}{\sqrt{N}}$ (Cf. ref.[125–127]), that is the fluctuation (or dispersion) in the number of particles $\sim \sqrt{N}$. This immediately leads to equations (2.1) and (2.2).
It must be emphasized that equations (2.1) and (2.2) in particular bring out apart from the random feature, a holistic or Machian feature in which the large scale universe and the micro world are inextricably tied up, as against the usual reductionist view discussed in detail earlier. This is in fact inescapable if we are to consider a Brownian Heap. This interpretation in which the extent $R (\text{or} \, T)$ in (2.1) (or (2.2)) is a dispersion also explains the fractal dimensionality 2: If the steps were laid out one beside the other unidirectionally as in conventional thinking, then we would have the usual

dimensionality one. For, instead of (2.1), we would have,

$$R = Nl$$

This again is tied up with a model in terms of a Weiner process (a Random Walk), as we will see below.

There is another nuance. Newtonian space was a passive container which "contained" matter and interactions—–these latter were actors performing on the fixed platform of space. But our view is in the spirit of Liebniz [128] for whom the container of space was made up of the contents—–the actors, as it were, made up the stage or platform. This also implies what is called background independence a feature shared by General Relativity (but not String Theory).

It should also be observed that the cut off length for fractal behaviour depends on the mass, via the de Broglie or Compton wavelength. The de Broglie wavelength is the non-relativistic version of the Compton wavelength. Indeed as has been shown in detail [129, 130], it is the zitterbewegung or self-interaction effects within the minimum cut off Compton wavelength that give rise to the inertial mass. So the appearance of mass in the minimum cut off Compton (or de Broglie) scale is quite natural. This point will be analyzed further in the sequel.

We can appreciate that the fractal nature and a stochastic underpinning are interrelated: for scales less than the Compton (or de Broglie) wavelength, time is irregular and can be modelled by a double Wiener process[131]. This will be shown to lead to the complex wave function of Quantum Mechanics, which is one of its distinguishing characteristics (in contrast to Classical theory where complex quantities are a mathematical artifice).

To appreciate all this let us consider the motion of a particle with position given by $x(t)$, subject to random correction given by, as in the usual theory, (Cf.[90, 108, 127]),

$$|\Delta x| = \sqrt{< \Delta x^2 >} \approx \nu \sqrt{\Delta t},$$

$$\nu = \hbar / m, \nu \approx lv \tag{2.3}$$

where ν is the so called diffusion constant and is related to the mean free path l as above. We can then proceed to deduce the Fokker-Planck equation as follows (Cf.ref.[90] for details):

We first define the forward and backward velocities corresponding to having time going forward and backward (or positive or negative time increments) in the usual manner,

$$\frac{d_+}{dt} x(t) = \mathbf{b}_+ \, , \, \frac{d_-}{dt} x(t) = \mathbf{b}_- \tag{2.4}$$

This leads to the Fokker-Planck equations

$$\partial \rho / \partial t + div(\rho \mathbf{b}_+) = V\Delta\rho,$$

$$\partial \rho / \partial t + div(\rho \mathbf{b}_-) = -U\Delta\rho \tag{2.5}$$

defining

$$V = \frac{\mathbf{b}_+ + \mathbf{b}_-}{2} \quad ; U = \frac{\mathbf{b}_+ - \mathbf{b}_-}{2} \tag{2.6}$$

We get on addition and subtraction of the equations in (2.5) the equations

$$\partial \rho / \partial t + div(\rho V) = 0 \tag{2.7}$$

$$U = \nu \nabla ln\rho \tag{2.8}$$

It must be mentioned that V and U are the statistical averages of the respective velocities. We can then introduce the definitions

$$V = 2\nu \nabla S \tag{2.9}$$

$$V - \imath U = -2\imath\nu\nabla(ln\psi) \tag{2.10}$$

We next observe the decomposition of the Schrodinger wave function as

$$\psi = \sqrt{\rho}e^{\imath S/\hbar}$$

leads to the well known Hamilton-Jacobi type equation

$$\frac{\partial S}{\partial t} = -\frac{1}{2m}(\partial S)^2 + \bar{\mathbf{V}} + Q, \tag{2.11}$$

where

$$Q = \frac{\hbar^2}{2m}\frac{\nabla^2\sqrt{\rho}}{\sqrt{\rho}}$$

From (2.9) and (2.10) we can finally deduce the usual Schrodinger equation or (2.11) [131].

We note that in this formulation three conditions are assumed, conditions whose import has not been clear. These are [90]:

(1) The current velocity is irrotational. Thus, there exists a function $S(x,t)$ such that

$$m\vec{V} = \vec{\nabla}S$$

(2) In spite of the fact that the particle is subject to random alterations in its motion there exists a conserved energy, defined in terms of its probability distribution.

(3) The diffusion constant is inversely proportional to the inertial mass of

the particle, with the constant of proportionality being a universal constant $\hbar$ (Cf. equation (2.3)):

$$\nu = \frac{\hbar}{m}$$

We note that the complex feature above disappears if the fractal or non-differential character is not present, (that is, the forward and backward time derivatives(2.6) are equal): Indeed the fractal dimension 2 also leads to the real coordinate becoming complex. What distinguishes Quantum Mechanics is the adhoc feature, the diffusion constant ν of (2.3) in Nelson's theory and the "Quantum potential" Q of (2.11) which appears in Bohm's theory [130] as well, though with a different meaning.
Interestingly from the Uncertainty Principle,

$$m\Delta x \frac{\Delta x}{\Delta t} \sim \hbar$$

we get back equation (2.3) of Brownian motion. This shows the close connection on the one hand, and provides, on the other hand, a rationale for the particular, otherwise adhoc identification of ν in (2.3)−−its being proportional to $\hbar$.
We would like to emphasize that we have arrived at the Quantum Mechanical Schrodinger equation from Classical considerations of diffusion, though with some new assumptions. In the above, effectively we have introduced a complex velocity $V - \imath U$ which alternatively means that the real coordinate x goes into a complex coordinate

$$x \to x + \imath x' \tag{2.12}$$

To see this in detail, let us rewrite (2.6) as

$$\frac{dX_r}{dt} = V, \quad \frac{dX_\imath}{dt} = U, \tag{2.13}$$

where we have introduced a complex coordinate X with real and imaginary parts X_r and $X_\imath$, while at the same time using derivatives with respect to time as in conventional theory.
We can now see from (2.6) and (2.13) that

$$W = \frac{d}{dt}(X_r - \imath X_\imath) \tag{2.14}$$

That is, in this non relativistic development either we use forward and backward time derivatives and the usual space coordinate as in (2.6), or we use the derivative with respect to the usual time coordinate but introduce complex space coordinates as in (2.12). Already, we can get a glimpse of the special relativistic hyperbolic geometry with real space and imaginary time coordinates (or vice versa).

Let us briefly analyze this aspect though we will return to it later. To bring out the new input here, we will consider the diffusion equation (2.3) in only one dimension for the moment. We note that through (2.6) we have introduced a complex velocity W, as indeed can be seen from (2.13) and (2.14) as well. Furthermore (2.8) and (2.9) show that both U and V can be written as gradients in the form

$$\vec{V} = \vec{\nabla} f$$

$$\vec{U} = \vec{\nabla} g \tag{2.15}$$

Furthermore the equation of continuity, (2.7) shows that for nearly constant and homogenous density ρ we have

$$\vec{\nabla} \cdot \vec{V} = 0 \tag{2.16}$$

where we are still retaining the vector notation. This implies that f and so also g satisfy the Laplacian equation

$$\nabla^2 f = 0 \tag{2.17}$$

In this case given (2.17),it is well known from the Theory of Fluid flow [132] that the trajectories $f =$ constant and $g =$ constant are orthogonal, with, in the case of spherical symmetry, the former representing radial stream lines and the latter circles around the origin (or more generally closed curves). We also see that (2.16) shows that the velocity is solenoidal, and $\vec{V}$ being a gradient, by (2.9), also irrotational. We would then expect that the circulation given by the expression

$$\Gamma = m \oint \vec{V} \cdot d\vec{s} \tag{2.18}$$

would vanish. All this is true in a simply connected space. However if the space is multiply connected, the origin being the singularity, then the circulation (2.18) does not vanish. We argue that this is the Quantum Mechanical spin, and will return to this point. But briefly, Γ in (2.18) equals the Quantum Mechanical spin $h/2$. This follows, if we take the radius of the circuit of integration to be the Compton wavelength $\hbar/mc$ and remember that at this distance, the velocity equals c.

The interesting thing is that starting from a single real coordinate, we have ended up with a complex coordinate, and have characterized thereby, the Quantum Mechanical spin. Indeed as we will shortly see it was noticed by

Newman in the derivation of the Kerr-Newman metric, that an inexplicable imaginary shift gives Quantum Mechanical spin. In other words Quantum Mechanics results from a complexification of coordinates, this as can be seen now, being symptomatic of multiply connected spaces, and modelled by the Weiner process above. We will return to this in more detail in later Chapters, particularly Chapters 5 and 8.

Finally, it may be remarked that the original Nelsonian theory itself has been criticized by different scholars [133]-[137].

To get further insight into the foregoing considerations, let us start with the Langevin equation in the absence of external forces,[108, 138]

$$m\frac{dv}{dt} = -\alpha v + F'(t)$$

where the coefficient of the frictional force is given by Stokes's Law [132]

$$\alpha = 6\pi\eta a$$

η being the coefficient of viscosity, and where we are considering a sphere of radius a. This then leads to two cases.

Case (i):

For t, there is a cut off time τ. It is known (Cf.[108]) that there is a characteristic time constant of the system, given by

$$\frac{m}{\alpha} \sim \frac{m}{\eta a},$$

so that, from Stokes's Law, as

$$\eta = \frac{mc}{a^2} \text{ or } m = \eta\frac{a^2}{c}$$

we get

$$\tau \sim \frac{ma^2}{mca} = \frac{a}{c},$$

that is τ is the Compton time.

The expression for η which follows from the fact that

$$F_x = \eta(\Delta s)\frac{dv}{dz} = m\dot{v} = \eta\frac{a^2}{c}\dot{v},$$

shows that the intertial mass is due to a type of "viscosity" of the background Zero Point Field (ZPF). (Cf. also ref.[139]). We will revisit this circumstance later on in Chapter 4.

To sum up case (i), if there is a cut off τ, the stochastic formulation leads us back to the minimum space time intervals $\sim$ Compton scale.

To push these small scale considerations further, we have, using the Beckenstein radiation equation[140],

$$t \equiv \tau = \frac{G^2 m^3}{\hbar c^4} = \frac{m}{\eta a} = \frac{a}{c}$$

which gives

$$a = \frac{\hbar}{mc} \quad \text{if} \quad \frac{Gm}{c^2} = a$$

In other words the Compton wavelength equals the Schwarzchild radius, which automatically gives us the Planck mass. Thus as noted the inertial mass is thrown up in these considerations. We will also see that the Planck mass leads to other particle masses.
On the other hand if we work with $t \geq \tau$ we get

$$ac = \frac{2kT}{\eta a}$$

whence

$$kT \sim mc^2,$$

which is the Hagedorn formula for Hadrons[141].
Thus both the Planck scale and the Compton wavelength Hadron scale considerations follow meaningfully.
Case (ii):
If there is no cut off time τ, as is known, we get back, equation (2.3),

$$\Delta x = \nu \sqrt{\Delta t}$$

and thence Nelson's derivation of the non relativistic Schrodinger equation. We can see here that the absence of a spacetime cut off leads to the non-relativistic theory, but on the contrary the cut off leads to the origin of the inertial mass (and as we will see, relativity itself). On the other hand, as we saw, the cut off is symptomatic of a multiply connected space- where we cannot shrink circuits to a point.
The relativistic generalization of the above considerations to the Klein-Gordon equation has been even more troublesome[71]. In this case, there are further puzzling features apart from the luminal velocity as in the Dirac equation. For Lorentz invariance, a discrete time is further required. Interestingly, as we will see Snyder had shown that discrete spacetime is compatible with Lorentz transformations. Here again, the Compton wavelength and time cut off will be seen to make the whole picture transparent. The stochastic derivation of the Dirac equation introduces a further complication. There is a spin reversal with the frequency $mc^2/\hbar$. This again is

readily explainable in the earlier context of zitterbewegung in terms of the Compton time. Interestingly the resemblance of such a Weiner process to the zitterbewegung of the electron was noticed by Ichinose[142].
Thus in all these cases once we recognize that the Compton wavelength and time are minimum cut off intervals, the obscure or adhoc features become meaningful.
We would like to reiterate that the origin of the Compton wavelength is the random walk equation (2.1)! One could then argue that the Compton time (or Chronon) automatically follows. This was shown by Hakim [77, 79]. Intutively, we can see that a discrete space would automatically imply discrete time. For, if Δt could $\to 0$, then all velocities, $lim_{\Delta t \to 0} |\frac{\Delta x}{\Delta t}|$ would $\to \infty$ as $|\Delta x|$ does not tend to 0! So there would be a minimum time cut off and a maximal velocity and this in conjunction with symmetry considerations can be taken to be the basis of special relativity as we will see below in more detail.
In fact one could show that quantized spacetime is more fundamental than quantized energy and indeed would lead to the latter. To put it simply the frequency is given by c/λ, where λ the wavelength is itself discrete and hence so also is the frequency. One could then deduce Planck's law as will be seen in the next Section (Cf.[14]). This of course, is the starting point of Quantum Theory itself.
At this stage we remark that in the case of the Dirac electron, the point electron has the velocity of light and is subject to zitterbewegung within the Compton wavelength. The thermal wavelength for such a motion is given by

$$\lambda = \sqrt{\frac{\hbar^2}{mkT}} \sim \text{De Broglie} \quad \text{wave length}$$

by virtue of the fact that now $kT \sim mv^2$ itself. In the limit $v \to c$ in the spirit of the luminal velocity of the point Dirac electron or, using the earlier relation, $kT \sim mc^2$, λ becomes the Compton wavelength. To look at this from another point of view, it is known that for a collection of relativistic particles, the various mass centres form a two-dimensional disc perpendicular to the angular momentum vector $\vec{L}$ and with radius (ref.[95])

$$r = \frac{L}{mc} \tag{2.19}$$

Further if the system has positive energies, then it must have an extension greater than r, while at distances of the order of r we begin to encounter negative energies.

If we consider the system to be a particle of spin or angular momentum $\frac{\hbar}{2}$, then equation (2.19) gives, $r = \frac{\hbar}{2mc}$. That is we get back the Compton wavelength. Another interesting feature which we will encounter later is the two dimensionality of the space or disc of mass centres.
On the other hand it is known that, if a Dirac particle is represented by a Gausssian packet, then we begin to encounter negative energies precisely at the same Compton wavelength as above. These considerations show the interface between Classical and Quantum considerations.
Infact as has been shown it is this circumstance that leads to inertial mass, while Gravitation and Electromagnetism (as for example brought out by the Kerr-Newman metric) and indeed QCD interactions also will be seen to follow in Chapter 6. In the light of the above remarks, it appears that the fractal or Brownian Heap character of space time is at the root of Quantum behaviour.

2.3 Spacetime

As remarked in the previous section, the fact that forward and backward time derivatives in the double Wiener process do not cancel leads to a complex velocity (cf.[131]), $V - \imath U$. That is, the usual space coordinate x (in one dimension for simplicity) is replaced by a coordinate like $x + \imath x'$, where x' is a non constant function of time that is, a new imaginary coordinate is introduced. We will now show that it is possible to consistently take $x' = ct$.
Let us take the simplest choice for x', viz., $x' = \lambda t$. Then the imaginary part of the complex velocity in (2.14)is given by $U = \lambda$. Then we have (cf.[127]),

$$U = \nu \frac{d}{dx}(ln\rho) = \lambda$$

where ν and ρ have been defined in (2.3), and in the equation leading to (2.11). We thus have, $\rho = e^{\gamma x}$, where $\gamma = \lambda/\nu$ and the quantum potential of (2.11) is given by

$$Q \sim \frac{\hbar^2}{2m} \cdot \quad \gamma^2 \tag{2.20}$$

In this stochastic formulation with Compton wavelength cut off, it is known that Q turns out to be the inertial energy mc^2. It then follows from (2.20) and the definition of γ, that $\lambda \approx c$.
In other words it is in the above stochastic formulation that we see the

emergence of the spacetime coordinates $(x, \imath ct)$ and Special Relativity from a Weiner process in which time is a back and forth process. All this has been in one dimension.
If we now generalize to three spatial dimensions, then as we will see in a moment [143], we get the quarternion formulation with the three Pauli spin matrices replacing $\imath$, giving the purely Quantum Mechanical spin half of Dirac. On the other hand, the above formulation with minimum space time cut offs will also be shown to lead independently to the Dirac equation. Thus the origin of Special Relativity, inertial mass and the Quantum Mechanical spin half is the minimum space time cut offs. We had already encountered this idea in Chapter 1.
We digress for a moment to observe that equations (2.1) and (2.2) indicate that the Compton scale is a fundamental unit of spacetime. We will now show that this quantized space time leads to Planck's quantized energy, as was briefly seen in the previous section.
The derivation is similar to the well known theory[144].
Let the energy be given by

$$E = g(\nu)$$

Then, f the average energy associated with each mode is given by,

$$f = \frac{\sum_\nu g(\nu)e^{-g(\nu)/kT}}{\sum_\nu e^{-g(v)/kT}}$$

Again, as in the usual theory, a comparison with Wien's functional relation, gives,

$$f = \nu F(\nu/kT),$$

whence,

$$E = g(\nu) \propto \nu,$$

which is Planck's law.
Yet another way of looking at it is, as the momentum and frequency of the classical oscillator have discrete spectra so does the energy.

2.4 Further Considerations

To see all this in greater detail, we observe that if we treat an electron as a Kerr-Newman Black Hole, then we get the correct Quantum Mechanical

$g = 2$ factor, but the horizon of the Black Hole becomes complex [130, 45].

$$r_+ = \frac{GM}{c^2} + \imath b, b \equiv \left(\frac{G^2M^2}{c^4} - \frac{GQ^2}{c^4} - a^2\right)^{1/2} \tag{2.21}$$

G being the gravitational constant, M the mass and $a \equiv L/Mc, L$ being the angular momentum. While (2.21) exhibits a naked singularity, and as such has no physical meaning, we note that from the realm of Quantum Mechanics the position coordinate for a Dirac particle in conventional theory is given by

$$x = (c^2 p_1 H^{-1} t) + \frac{\imath}{2} c\hbar(\alpha_1 - cp_1 H^{-1})H^{-1} \tag{2.22}$$

an expression that is very similar to (2.21). Infact as was argued in detail [130] the imaginary parts of both (2.21) and (2.22) are the same, being of the order of the Compton wavelength.

It is at this stage that a proper physical interpretation begins to emerge. Dirac himself observed as noted, that to interpret (2.22) meaningfully, it must be remembered that Quantum Mechanical measurements are really averaged over the Compton scale: Within the scale there are the unphysical zitterbewegung effects: for a point electron the velocity equals that of light. Once such a minimum spacetime scale is invoked, then we have a non commutative geometry as shown by Snyder more than fifty years ago [145, 146]:

$$[x, y] = (\imath a^2/\hbar)L_z, [t, x] = (\imath a^2/\hbar c)M_x, etc.$$

$$[x, p_x] = \imath\hbar[1 + (a/\hbar)^2 p_x^2]; \tag{2.23}$$

The relations (2.23) are compatible with Special Relativity. Indeed such minimum spacetime models were studied for several decades, precisely to overcome the divergences encountered in Quantum Field Theory [130],[146]-[151], [152, 153].

Before proceeding further, it may be remarked that when the square of a, which we will take to be the Compton wavelength (including the Planck scale, which is a special case of the Compton scale for a Planck mass viz., $10^{-5}gm$), in view of the above comments can be neglected, then we return to point Quantum Theory.

It is interesting that starting from the Dirac coordinate in (2.22), we can deduce the non commutative geometry (2.23), independently. For this we note that the α's in (2.22) are given by

$$\vec{\alpha} = \begin{bmatrix} \vec{\sigma} & 0 \\ 0 & \vec{\sigma} \end{bmatrix},$$

the σ's being the Pauli matrices. We next observe that the first term on the right hand side is the usual Hermitian position. For the second term which contains α, we can easily verify from the commutation relations of the σ's that

$$[x_\imath, x_\jmath] = \beta_{\imath\jmath} \cdot l^2 \tag{2.24}$$

where l is the Compton scale.

There is another way of looking at this. Let us consider the one dimensional coordinate in (2.22) or (2.21) to be complex. We now try to generalize this complex coordinate to three dimensions. Then as briefly noted, in the previous Section, we encounter a surprise——we end up with not three, but four dimensions,

$$(1, \imath) \to (I, \sigma),$$

where I is the unit 2×2 matrix and σs are the Pauli matrices. We get the special relativistic Lorentz invariant metric at the same time. (In this sense, as noted by Sachs [143], Hamilton who made this generalization would have hit upon Special Relativity, if he had identified the new fourth coordinate with time).

That is,

$$x + \imath y \to Ix_1 + \imath x_2 + \jmath x_3 + kx_4,$$

where $(\imath, \jmath, k)$ now represent the Pauli matrices; and, further,

$$x_1^2 + x_2^2 + x_3^2 - x_4^2$$

is invariant. Before proceeding further, we remark that special relativistic time emerges above from the generalization of the complex one dimensional space coordinate to three dimensions, just as the relativistic time came out of the one dimensional space coordinate as seen earlier.

While the usual Minkowski four vector transforms as the basis of the four dimensional representation of the Poincare group, the two dimensional representation of the same group, given by the right hand side in terms of Pauli matrices, obeys the quaternionic algebra of the second rank spinors (Cf.Ref.[154, 155, 143] for details).

To put it briefly, the quarternion number field obeys the group property and this leads to a number system of quadruplets as a minimum extension. In fact one representation of the two dimensional form of the quarternion basis elements is the set of Pauli matrices. Thus a quarternion may be expressed in the form

$$Q = -\imath\sigma_\mu x^\mu = \sigma_0 x^4 - \imath\sigma_1 x^1 - \imath\sigma_2 x^2 - \imath\sigma_3 x^3 = (\sigma_0 x^4 + \imath\vec{\sigma} \cdot \vec{r})$$

This can also be written as

$$Q = -\imath \begin{pmatrix} \imath x^4 + x^3 & x^1 - \imath x^2 \\ x^1 + \imath x^2 & \imath x^4 - x^3 \end{pmatrix}.$$

As can be seen from the above, there is a one to one correspondence between a Minkowski four-vector and Q. The invariant is now given by $Q\bar{Q}$, where $\bar{Q}$ is the complex conjugate of Q.
However, as is well known, there is a lack of spacetime reflection symmetry in this latter formulation. If we require reflection symmetry also, we have to consider the four dimensional representation,

$$(I, \vec{\sigma}) \to \left[\begin{pmatrix} I & 0 \\ 0 & -I \end{pmatrix}, \begin{pmatrix} 0 & \vec{\sigma} \\ \vec{\sigma} & 0 \end{pmatrix} \right] \equiv (\Gamma^\mu)$$

(Cf.also.ref. [156] for a detailed discussion). The motivation for such a reflection symmetry is that usual laws of physics, like electromagnetism do indeed show the symmetry.
We at once deduce spin and Special Relativity and the geometry (2.23) in these considerations. This is a transition that has been long overlooked [157, 158]. It must also be mentioned that spin half itself is relational and refers to three dimensions, to a spin network infact [159, 45]. That is, spin half is not meaningful in a single particle Universe.
While a relation like (2.24) above has been in use recently, in non commutative models, we would like to stress that it has been overlooked that the origin of this non commutativity lies in the original Dirac coordinates.
The above relation shows on comparison with the position-momentum commutator that the coordinate $\vec{x}$ also behaves like a "momentum". This can be seen directly from the Dirac theory itself where we have [75]

$$c\vec{\alpha} = \frac{c^2\vec{p}}{H} - \frac{2\imath}{\hbar}\hat{x}H \tag{2.25}$$

In (2.25), the first term is the usual momentum. The second term is the extra "momentum" $\vec{p}$ due to zitterbewegung.
Infact we can easily verify from (2.25) that

$$\vec{p} = \frac{H^2}{\hbar c^2}\hat{x} \tag{2.26}$$

where $\hat{x}$ has been defined in (2.25).
We finally investigate what the angular momentum $\sim \vec{x} \times \vec{p}$ gives—–that is, the angular momentum at the Compton scale. We can easily show that

$$(\vec{x} \times \vec{p})_z = \frac{c}{E}(\vec{\alpha} \times \vec{p})_z = \frac{c}{E}(p_2\alpha_1 - p_1\alpha_2) \tag{2.27}$$

where E is the eigen value of the Hamiltonian operator H. Equation (2.27) shows that the usual angular momentum but in the context of the minimum Compton scale cut off, leads to the "mysterious" Quantum Mechanical spin.
In the above considerations, we started with the Dirac equation and deduced the underlying non commutative geometry of spacetime. Interestingly, starting with Snyder's non commutative geometry, based solely on Lorentz invariance and a minimum spacetime length, which we have taken to be the Compton scale, (2.23), it is possible to deduce the relations (2.27), (2.26) and the Dirac equation itself as we will see later.
We have thus established the correspondence between considerations starting from the Dirac theory of the electron and Snyder's (and subsequent) approaches based on a minimum spacetime interval and Lorentz covariance. It has been argued in Chapter 1 from an alternative point of view that Special Relativity operates outside the Compton wavelength.
We started with the Kerr-Newman Black Hole. Infact the derivation of the Kerr-Newman Black Hole itself begins with a Quantum Mechanical spin yielding complex shift, which Newman has found inexplicable even after several decades [160, 161]. As he observed, "...one does not understand why it works. After many years of study I have come to the conclusion that it works simply by accident". And again, "Notice that the magnetic moment $\mu = ea$ can be thought of as the imaginary part of the charge times the displacement of the charge into the complex region... We can think of the source as having a complex center of charge and that the magnetic moment is the moment of charge about the center of charge... In other words the total complex angular momentum vanishes around any point z^a on the complex world-line. From this complex point of view the spin angular momentum is identical to orbital, arising from an imaginary shift of origin rather than a real one... If one again considers the particle to be "localized" in the sense that the complex center of charge coincides with the complex center of mass, one again obtains the Dirac gyromagnetic ratio..."
The unanswered question has been, why does a complex shift in classical theory somehow represent Quantum Mechanical spin about that axis? The question has now been answered. Complexified spacetime is symptomatic of fuzzy spacetime and a non commutative geometry and Quantum Mechanical spin and relativity. Indeed Zakrzewski has shown in a classical context that non commutativity implies spin [162, 163]. We will return to these considerations later.
The above considerations recovered the Quantum Mechanical spin together

with classical relativity, though the price to pay for this was minimum spacetime intervals and noncommutative geometry.

2.5 The Path Integral Formulation

We come to another description of Quantum Mechanics and first argue that the alternative Feynman Path Integral formulation essentially throws up fuzzy spacetime. To recapitulate [122, 164, 165], if a path is given by

$$x = x(t)$$

then the probability amplitude is given by

$$\phi(x) = e^{\imath \int_{t_1}^{t_2} L(x,\dot{x})dt}$$

So the total probability amplitude is given by

$$\sum_{x(t)} \phi(x) = \sum e^{\imath \int_{t_1}^{t_2} L(x,\dot{x})dt} \equiv \sum e^{\frac{\imath}{\hbar} S}$$

In the Feynman analysis, the path

$$x = \bar{x}(t)$$

appears as the actual path for which the action is stationery. From a physical point of view, for paths very close to this, there is constructive interference, whereas for paths away from this the interference is destructive. We will see later that this is in the spirit of the formulation of the random phase. However it is well known that the convergence of the integrals requires the Lipshitz condition viz.,

$$\Delta x^2 \approx a\Delta t \tag{2.28}$$

We could say that only those paths satisfying (2.28) constructively interfere. We would now like to observe that (2.28) is the same as the Brownian or Diffusion equation (2.3) related to our earlier discussion of the Weiner process. The point is that (2.28) again implies a minimum spacetime cut off, as indeed was noted by Feynman himself [122], for if Δt could $\to 0$, then the velocity would $\to \infty$.

To put it another way we are taking averages over an interval Δt, within which there are unphysical processes as noted. It is only after the average is taken, that we recover physical spacetime intervals which hide the fractality or unphysical feature. If in the above, Δt is taken as the Compton time and a is identified with the earlier ν, then we recover for the root mean

squared velocity, the velocity of light.
As we have argued in detail this is exactly the situation which we encounter in the Dirac theory of the electron. There we have the unphysical zitterbewegung effects within the Compton time Δt and as $\Delta t \to 0$ the velocity of the electron tends to the maximum possible velocity, that of light [15].
This existence of a minimum spacetime scale, it has been argued is the origin of fuzzy spacetime, described by a noncommutative geometry, consistent with Lorentz invariance viz., equations (2.23) and (2.24).
We reiterate that the momentum position commutation relations lead to the usual Quantum Mechanical commutation relations in the usual (commutative) spacetime if $O(l^2)$ is neglected where l defines the minimum scale. Indeed, we have at the smallest scale, a quantum of area reflecting the fractal dimension, the Quantum Mechanical path having the fractal dimension 2 (Cf.ref.[111]). It is this "fine structure" of spacetime which is expressed in the noncommutative structure (2.23) or (2.24). Neglecting $O(l^2)$ is equivalent to neglecting the above and returning to usual spacetime. In other words Snyder's purely classical considerations at a Compton scale lead to Quantum Mechanics as will be discussed in Chapter 6.
In the light of the above comments, we can now notice that within the Compton time, we have a double Weiner process leading to non differentiability with respect to time. That is, at this level time in our usual sense does not exist. To put it another way, within the Compton scale we have the complex or non-Hermitian position coordinates for the Dirac electron and zitterbewegung effects−−as we saw, these are unphysical, non local and chaotic in a literal sense.
This is a Quantum Mechanical and an experimental fact. It expresses the Heisenberg Uncertainty Principle−−spacetime points imply infinite momenta and energies. However as noted earlier Quantum Theory has lived with this contradiction. To reiterate to measure space or time intervals we need units which can be to a certain extent and not indefinitely subdivided−−but already this is the origin of discreteness. That is, our measurements are resolution dependent. So physical time emerges at values greater than the minimum unit, which has been shown to be at the Compton scale. Going to the limit of space-time points leads to the well known infinities of Quantum Field Theory (and classical electron theory) which require renormalization for their removal.
The conceptual point here is that time is in a sense synonymous with change, but this change has to be tractable or physical. The non differentiability with respect to time, symbolized and modeled by the double Weiner

process, within the Compton time, precisely highlights time or change which is not tractable, that is, is unphysical. However Physics, tractability and differentiability emerge from this indeterminism once averages over the zitterbewegung or Compton scale are taken. It is now possible to track time physically in terms of multiples of the Compton scale.

2.6 Remarks

1. We would like to make the following observations:
i) We have in effect equated the statistical fluctuations, when there are N particles to the Quantum Mechanical fluctuations. The former fluctuations take place over a scale $\sim R/\sqrt{N}$, where R is the size of the system of particles and N is the number of particles in the system. The Quantum Mechanical fluctuations take place at a scale of the order of the Compton wavelength as we will see in the next Chapter. Apart from the fact that the equality of these two has been taken to be an empirical coincidence, we actually deduce this equality in our cosmology in the next Chapter. Thus the equality is no longer accidental or ad hoc. However a nuance must be borne in mind. In the conventional theory, the Quantum Mechanical fluctuation is a reductionist effect, whereas the statistical fluctuation is a "thermodynamic" or statistical effect in a collection of particles.
We may further add, in this context, that it is possible to arrive at the Hamilton-Jacobi equation (2.11) and thence Quantum Mechanics, by considering the fluctuations in the General Relativistic metric from a cosmological point of view. This will be examined briefly in the next two Chapters.
ii) In the random mechanical approach, including Nelson's, we encounter the "potential" $Q-$ this represents in the usual theory a peculiar correlation between the random motion of a particle and its probability distribution function.
iii) We would like to point out that it would be reasonable to expect that the Weiner process discussed earlier is related to the ZPF which is the Zero Point Energy of a Quantum Harmonic oscillator. We can justify this expectation as follows: Let us denote the forward and backward time derivatives as before by d_+ and d_-. In usual theory where time is differentiable, these two are equal, but we have on the contrary taken them to be unequal. Let

$$d_- = a - d_+ \tag{2.29}$$

Then we have from Newton's second law in the absence of forces,

$$\ddot{x} + k^2 x = a\dot{x} \tag{2.30}$$

wherein the new nondifferentiable effect (2.29) is brought up. In a normal vacuum with usual derivatives and no external forces, Newtonian Mechanics would give us instead the equation

$$\ddot{x} = 0 \tag{2.31}$$

A comparison of (2.30) and (2.31) shows that the Weiner process converts a uniformly moving particle, or a particle at rest into an oscillator. Indeed in (2.30) if we take as a first approximation

$$\dot{x} \approx \langle \dot{x} \rangle = 0 \tag{2.32}$$

then we would get the exact oscillator equation

$$\ddot{x} + k^2 x = 0 \tag{2.33}$$

for which in any case, consistently (2.32) is correct. We can push these considerations even further and deduce alternatively, the Schrodinger equation, as seen earlier. The genesis of Special Relativity too can be found in the Weiner process. Let us examine this more closely.

We first define a complete set of base states by the subscript $\imath$ and $U(t_2, t_1)$ the time elapse operator that denotes the passage of time between instants t_1 and t_2, t_2 greater than t_1. We denote by, $C_\imath(t) \equiv < \imath|\psi(t) >$, the amplitude for the state $|\psi(t) >$ to be in the state $|\imath >$ at time t, and [129, 130]

$$< \imath|U|j > \equiv U_{\imath j}, U_{\imath j}(t + \Delta t, t) \equiv \delta_{\imath j} - \frac{\imath}{\hbar} H_{\imath j}(t)\Delta t.$$

We can now deduce from the super position of states principle that,

$$C_\imath(t + \Delta t) = \sum_j [\delta_{\imath j} - \frac{\imath}{\hbar} H_{\imath j}(t)\Delta t] C_j(t) \tag{2.34}$$

and finally, in the limit,

$$\imath\hbar \frac{dC_\imath(t)}{dt} = \sum_j H_{\imath j}(t) C_j(t) \tag{2.35}$$

where the matrix $H_{\imath j}(t)$ is identified with the Hamiltonian operator. We have argued earlier at length that (2.35) leads to the Schrodinger equation [129, 130]. In the above we have taken the usual unidirectional time to deduce the non relativistic Schrodinger equation. If however we consider a Weiner process in (2.34) then we will have to consider instead of (2.35)

$$C_\imath(t - \Delta t) - C_\imath(t + \Delta t) = \sum_j \left[\delta_{\imath j} - \frac{\imath}{\hbar} H_{\imath j}(t)\Delta t\right] C_j^{(t)} \qquad (2.36)$$

Equation (2.36) in the limit can be seen to lead to the relativistic Klein-Gordon equation rather than the Schrodinger equation [166]. This is an alternative justification for our earlier result that Special Relativity emerges from the above considerations. Furthermore, the Klein-Gordon equation describes the normal mode vibrations of Harmonic Oscillators−−that is, we recover (2.33), again.

2. We have seen that the path integral formulation is an alternative to the Schrodinger equation, an alternative that has a resemblance to the stochastic mechanics encountered earlier. However we should bear in mind that these paths are merely mathematical tools for computing the evolution of the wave functions [167]. Nevertheless we should note that the path integral formulation does not give the probability distribution on the space of all paths, so that we cannot legitimately conclude that nature chooses one of the several paths at random according to the probability distribution. Unfortunately in this formulation the measures is complex (and not even rigorously defined in the limit of the continuum). Nor will the imaginary or real paths of the measure give the actual Quantum Mechanical picture. It would be more correct to say that the paths are possible paths for a part of the Quantum Mechanical wave. In any case, all this reflects via (2.28), the unphysicality within the minimum interval Δt.

On the other hand there is the well known Bohmian formulation of Quantum Mechanics which uses the Schrodinger wave function, and the Schrodinger equation to deduce the Hamilton-Jacobi equation exactly as in the stochastic case. But the resemblance is superficial. This non relativistic formulation is one in which the observer plays no part. There is a hidden variable in the form of the position coordinate of the particle. Thus one of the Bohmian paths represents the actual motion of the particle, which exists separately from the wave function. Moreover the Quantum potential Q in the Bohmian case has a non local character and no clear explanation. Furthermore there is no clear generalization to the relativistic case. For all these reasons, though Bohm studied this approach in the 1950s, it has not really caught on and we will not pursue the matter further.

3. As mentioned discrete spacetime and some of their effects have been studied from different points of view for several decades now. It is worth

mentioning here that the usual notion of time as an operator with continuous eigen values in Quantum Theory runs into difficulty, as was appreciated by Pauli a long time ago[168]. This can be seen by a simple argument, and, we follow Park [169]: Let the time operator be denoted by $\hat{T}$, satisfying

$$\left[\hat{T}, \hat{H}\right] = \imath.$$

Let $|E'>$ be an eigenfunction of $\hat{H}$ belonging to the eigenvalue E', and let $|E'>_\epsilon = e^{\imath\epsilon\hat{T}}|E'>$. Then

$$\hat{H}|E'>_\epsilon = e^{\imath\epsilon\hat{T}} e^{-\imath\epsilon\hat{T}} \hat{H} e^{\imath\epsilon\hat{T}}|E'> = (E' + \epsilon)|E'>) \qquad (2.37)$$

Remembering that ϵ is arbitrary, (2.37) gives a continuous energy spectrum, contrary to Quantum Theory. The difficulty is resolved if in the above considerations time were discrete.

4. It must be emphasized that in the stochastic formulation given in this Chapter, there are no hidden variables as in the Bohm formulation, due to the randomness or stochasticity, itself[121].

5. Though we will return to some of the above considerations later, it must be re-emphasized that in the absence of the double Weiner process alluded to, the imaginary part of the complex velocity potential U, vanishes, that is, so does ν of equation (2.3). In this case we come back to the domain of classical non-relativistic physics. So the origin of special relativity and Quantum Mechanics is to be found here in this double Weiner process within the Compton scale [16]. As pointed out in [130] non-relativistic Quantum Mechanics is not really compatible with Galilean or Newtonian Mechanics.

6. Finally, we would like to reemphasize the following point: By neglecting terms of the order l^2 (the squared Compton length) but not l itself, we return to point, commutative space time and can still have Quantum Mechanics and even relativistic Quantum Mechanics and Quantum Field Theory, though we would then have to introduce Quantum Mechanical spin by separate arguments and consider averages over the Compton scale anyway. But in the process, we are neglecting the Quantum of area or Abbot and Wise's fractal dimension of the Quantum Mechanical path. That is, we are snuffing out the fine structure implied by Quantum Theory and are then using, as remarked earlier, a thick brush to fudge. A quick way to see the result of Abbot and Wise is as follows [121]. From (2.3) it follows that

$$\langle v^2 \rangle \propto (\Delta t)^{-1}$$

Now if the Hausdorf dimension [112] is D, we would have,

$$\Delta t = (\Delta x)^D$$

whence

$$\langle v^2 \rangle \propto (\Delta t)^{2[(\frac{1}{D})-1]}$$

A comparison yields, $D = 2$.

Chapter 3

The Universe of Fluctuations

3.1 The New Cosmos

It may appear paradoxical, but the next step in our considerations of the microscopic structure of spacetime would be to consider the universe at large. Indeed this was a cue we could get from (2.1) and (2.2) of the previous Chapter itself. In the last century, we inherited the Newtonian Universe which was one in which there was an absolute background space in which the basic building blocks of the Universe were strewn about−−these were stars.

When Einstein proposed his General Theory of Relativity early in the last century, the accepted picture of the Universe was one where all major constituents were stationary. This had puzzled Einstein, because the gravitational pull of these constituents should make the Universe collapse as the nett force would be directed inwards. So he invented his famous cosmological constant, essentially a repulsive force that would counterbalance the attractive gravitational force.

Shortly thereafter as we saw briefly in Chapter 1, two dramatic discoveries completely transformed that picture. The first was due to astronomer Edwin Hubble, who discovered that the basic constituents or building blocks of the Universe were not stars, but rather, huge conglomerations of stars, called galaxies. The second discovery, aided by the redshift observations of the light of the galaxies was the fact that these galaxies are rushing away from each other. Rather than being static, the Universe is exploding. There was no need for the counterbalancing cosmic repulsion any more and Einstein dismissed his proposal as his greatest blunder.

By the end of the last century, the Big Bang Model had been worked out. It contained a huge amount of unobserved, hypothesized "matter" of a new kind−−dark matter. This was postulated as long back as the 1930s to explain the fact that the velocity curves of the stars in the galaxies did not fall off, as they should. Instead they flattened out, suggesting that the galaxies contained some undetected and therefore non-luminous or dark matter. The identity of this dark matter has been a matter of guess work, though. It could consist of Weakly Interacting Massive Particles (WIMPS) or Super Symmetric partners of existing particles. Or heavy neutrinos or monopoles or unobserved brown dwarf stars and so on. In fact Prof. Abdus Salam speculated some two decades ago [170] "And now we come upon the question of dark matter which is one of the open problems of cosmology. This is a problem which was speculated upon by Zwicky fifty years ago. He showed that visible matter of the mass of the galaxies in the Coma cluster was inadequate to keep the galactic cluster bound. Oort claimed that the mass necessary to keep our own galaxy together was at least three times that concentrated into observable stars. And this in turn has emerged as a central problem of cosmology.

"You see there is the matter which we see in our galaxy. This is what we suspect from the spiral character of the galaxy keeping it together. And there is dark matter which is not seen at all by any means whatsoever. Now the question is what does the dark matter consist of? This is what we suspect should be there to keep the galaxy bound. And so three times the mass of the matter here in our galaxy should be around in the form of the invisible matter. This is one of the speculations."

The universe in this picture, contained enough of the mysterious dark matter to halt the expansion and eventually trigger the next collapse. It must be mentioned that the latest WMAP survey [171], in a model dependent result indicates that as much as twenty three percent of the Universe is made up of dark matter, though there is no definite observational confirmation of its existence.

That is, the Universe would expand up to a point and then collapse.

There still were several subtler problems to be addressed. One was the famous horizon problem. To put it simply, the Big Bang was an uncontrolled or random event and so, different parts of the Universe in different directions were disconnected at the very earliest stage and even today, light would not have had enough time to connect them. So they need not be the same. Observation however shows that the Universe is by and large uniform, rather like people in different countries showing the same habits or dress. That would not be possible without some form of faster than light intercommunication which would violate Einstein's Special Theory of Relativity.

The next problem was that according to Einstein, due to the material content in the Universe, space should be curved whereas the Universe appears to be flat. There were other problems as well. For example astronomers predicted that there should be monopoles that is, simply put, either only North magnetic poles or only South magnetic poles, unlike the North South combined magnetic poles we encounter. Such monopoles have failed to show up even after seventy five years.

Some of these problems as we noted, were sought to be explained by what has been called inflationary cosmology whereby, early on, just after the Big Bang the explosion was super fast [46, 172].

What would happen in this case is, that different parts of the Universe, which could not be accessible by light, would now get connected. At the same time, the super fast expansion in the initial stages would smoothen out any distortion or curvature effects in space, leading to a flat Universe and in the process also eliminate the monopoles.

Nevertheless, inflation theory has its problems. It does not seem to explain the cosmological constant observed since. Further, this theory seems to imply that the fluctuations it produces should continue to indefinite distances. Observation seems to imply the contrary.

One other feature that has been studied in detail over the past few decades is that of structure formation in the Universe. To put it simply, why is the Universe not a uniform spread of matter and radiation? On the contrary it is very lumpy with planets, stars, galaxies and so on, with a lot of space separating these objects. This has been explained in terms of fluctuations in density, that is, accidentally more matter being present in a given region. Gravitation would then draw in even more matter and so on. These fluctuations would also cause the cosmic background radiation to be non uniform or anisotropic. Such anisotropies are in fact being observed. But this is not the end of the story. The galaxies seem to be arranged along two dimensional structures and filaments with huge separating voids. We will return to this very curious feature in the last Chapter.

From 1997, the conventional wisdom of cosmology that had concretized from the mid sixties onwards, began to be challenged. It had been believed that the density of the Universe is near its critical value, separating eternal expansion and ultimate contraction, while the nuances of the dark matter theories were being fine tuned. But that year, the author proposed a contra view, which we will examine in this Chapter. To put it briefly, the universe is permeated by a background dark energy, the Quantum Zero Point Field.

There would be fluctuations in this all permeating Zero Point Field—–or dark energy in the process of which, particles would be created [173–176]. This model while consistent with astrophysical observations predicted an ever expanding and accelerating Universe with a small cosmological constant. It deduces from theory the so called Large Number coincidences including the purely empirical Weinberg formula that connects the pion mass to the Hubble Constant [43, 17]—–"coincidences" that have troubled and mystified scientists from time to time.
However the work of Perlmutter and others [177, 178] began appearing in 1998 and told a different story. These observations of distant supernovae indicated that contrary to widely held belief, the Universe was not only not decelerating, it was actually accelerating though slowly. All this was greeted by the community with skepticism—–either it was plain wrong, or, let us wait and see.
A 2000 article in the Scientific American [179] observed, "In recent years the field of cosmology has gone through a radical upheaval. New discoveries have challenged long held theories about the evolution of the Universe... Now that observers have made a strong case for cosmic acceleration, theorists must explain it.... If the recent turmoil is anything to go by, we had better keep our options open."
On the other hand, an article in Physics World in the same year noted [180], "A revolution is taking place in cosmology. New ideas are usurping traditional notions about the composition of the Universe, the relationship between geometry and destiny, and Einstein's greatest blunder."
The infamous cosmological constant was resurrected and now it was "dark energy" that was in the air, rather than dark matter. The universe had taken a U turn.
Let us now examine this new cosmology and some of its implications. We will first go over the essentials and then examine the nuances.

3.2 The Mysterious Dark Energy

We first observe that the concept of a Zero Point Field (ZPF) or Quantum vacuum (or Aether) is an idea whose origin can be traced back to Max Planck himself. Quantum Field Theory attributes the ZPF to the virtual Quantum effects of an already present electromagnetic field [181]. What is the mysterious energy of supposedly empty vacuum? [182].
It may sound contradictory to attribute energy or density to the vacuum.

After all vacuum in the older concept is a total void. However, over the past four hundred years, it has been realized that it may be necessary to replace the vacuum by a medium with some specific physical properties. These properties were chosen to suit the specific requirements of the time. For instance Descartes the seventeenth century French philosopher mathematician proclaimed that the so called empty space above the mercury column in a Torricelli tube, that is, what is called the Torricelli vacuum, is not a vacuum at all. Rather, he said, it was something which was neither mercury nor air, something he called aether.

The seventeenth century Dutch Physicist, Christian Huygens required such a non intrusive medium like aether, so that light waves could propagate through it, rather like the ripples on the surface of a pond. This was the luminiferous aether. In the nineteenth century the aether was reinvoked. Firstly in a very intuitive way Faraday could conceive of magnetic effects in vacuum in connection with his experiments on induction. Based on this, the aether was used for the propagation of electromagnetic waves in Maxwell's Theory of electromagnetism, which in fact laid the stage for Special Relativity. This aether was a homogenous, invariable, non-intrusive, material medium which could be used as an absolute frame of reference at least for certain chosen observers. The experiments of Michelson and Morley towards the end of the nineteenth century were sought to be explained in terms of aether that was dragged by bodies moving in it. Such explanations were untenable and eventually lead to its downfall, and thus was born Einstein's Special Theory of Relativity in which there is no such absolute frame of reference. The aether lay discarded once again.

Very shortly thereafter the advent of Quantum Mechanics lead to its rebirth in a new and unexpected avatar. Essentially there were two new ingredients in what is today called the Quantum vacuum. The first was a realization that Classical Physics had allowed an assumption to slip in unnoticed: In a source or charge free "vacuum", one solution of Maxwell's Equations of electromagnetic radiation is no doubt the zero solution. But there is also a more realistic non zero solution. That is, the electromagnetic radiation does not necessarily vanish in empty space.

The second ingredient was the mysterious prescription of Quantum Mechanics, the Heisenberg Uncertainty Principle, according to which it would be impossible to precisely assign momentum and energy on the one hand and spacetime location on the other. Clearly the location of a vacuum with no energy or momentum cannot be specified in spacetime.

This leads to what is called a Zero Point Field. For instance a Harmonic oscillator, a swinging pendulum for example, according to classical ideas has zero energy and momentum in its lowest position. But the Heisenberg Uncertainty endows it with a fluctuating energy. This fact was recognized by Einstein himself way back in 1913, who contrary to popular belief, retained the concept of aether though from a different perspective [183]. It also provides an understanding of the fluctuating electromagnetic field in vacuum. Indeed, we have already seen in the previous Chapter, that this can be modeled by a Weiner process.

From another point of view, according to classical ideas, at the absolute zero of temperature, there should not be any motion. After all the zero is when all thermodynamic motion ceases. But as Nernst, father of the third law of Thermodynamics himself noted, experimentally this is not so. There is the well known superfluidity due to Quantum Mechanical−−and not thermodynamic−−effects. This is the situation where supercooled Helium moves in a spooky fashion.

This mysterious Zero Point Field or Quantum vacuum energy has since been experimentally confirmed in effects like the Casimir effect which demonstrates a force between uncharged parallel plates separated by a charge free medium, the Lamb shift which demonstrates a minute jiggling of an electron orbiting the nucleus in an atom−−as if it was being buffeted by the Zero Point Field, and as we will see, the anomalous Quantum Mechanical gyromagnetic ratio $g = 2$, the Quantum Mechanical spin half and so on [184]-[186], [45].

The Quantum vacuum is a far cry however, from the passive aether of olden days. It is a violent medium in which charged particles like electrons and positrons are constantly being created and destroyed, almost instantly, in fact within the limits permitted by the Heisenberg Uncertainty Principle for the violation of energy conservation. One might call the Quantum vacuum as a new state of matter, a compromise between something and nothingness. Something which corresponds to what the Rig Veda described thousands of years ago: "Neither existence, nor non existence."

Quantum vacuum can be considered to be the lowest state of any Quantum field, having zero momentum and zero energy. The fluctuating energy or ZPF due to Heisenberg's principle has an infinite value and is "renormalized", that is, discarded. The properties of the Quantum vacuum can under certain conditions be altered, which was not the case with the erstwhile aether. In modern Particle Physics, the Quantum vacuum is responsible for apart from effects alluded to earlier, other phenomena like quark confinement, a property we already encountered, whereby it would be impossible to observe an independent or free quark, the spontaneous breaking of symmetry of the electro weak theory, vacuum polarization wherein charges like electrons are surrounded by a cloud of other opposite charges tending to mask the main charge and so on. There could be regions of vacuum fluctuations comparable to the domain structures of ferromagnets. In a ferromagnet, all elementary electron-magnets are aligned with their spins in a certain direction. However there could be special regions wherein the spins are aligned differently.

Such a Quantum vacuum can be a source of cosmic repulsion, as pointed by Zeldovich and others [187, 130]. However a difficulty in this approach has been that the value of the cosmological constant turns out to be huge, far beyond what is observed. This has been called the cosmological constant problem [188].

There is another approach, that we briefly encountered in the previous Chapter, Stochastic Electrodynamics which treats the ZPF as independent and primary and attributes to it Quantum Mechanical effects [82, 92]. It may be re-emphasized that the ZPF results in the well known experimentally verified Casimir effect [189, 190]. We would also like to point out that contrary to popular belief, the concept of aether has survived over the decades through the works of Dirac, Vigier, Prigogine, String Theorists like Wilzeck and others [68],[191]-[195]. As pointed out it appears that even Einstein himself continued to believe in this concept [196].

We would first like to observe that the energy of the fluctuations in the background electromagnetic field could lead to the formation of elementary particles. Indeed this was Einstein's belief. As Wilzeck (loc.cit) put it, "Einstein was not satisfied with the dualism. He wanted to regard the fields, or ethers, as primary. In his later work, he tried to find a unified field theory, in which electrons (and of course protons, and all other particles) would emerge as solutions in which energy was especially concentrated, perhaps as singularities. But his efforts in this direction did not lead to any tangible success."

We will now argue that indeed this can happen. In the words of Wheeler [45], "From the zero-point fluctuations of a single oscillator to the fluctuations of the electromagnetic field to geometrodynamic fluctuations is a natural order of progression..."

Let us consider, following Wheeler a harmonic oscillator in its ground state. The probability amplitude is

$$\psi(x) = \left(\frac{m\omega}{\pi\hbar}\right)^{1/4} e^{-(m\omega/2\hbar)x^2}$$

for displacement by the distance x from its position of classical equilibrium. So the oscillator fluctuates over an interval

$$\Delta x \sim (\hbar/m\omega)^{1/2}$$

The electromagnetic field is an infinite collection of independent oscillators, with amplitudes X_1, X_2 etc. The probability for the various oscillators to have amplitudes X_1, X_2 and so on is the product of individual oscillator amplitudes:

$$\psi(X_1, X_2, \cdots) = exp[-(X_1^2 + X_2^2 + \cdots)]$$

wherein there would be a suitable normalization factor. This expression gives the probability amplitude ψ for a configuration $B(x, y, z)$ of the magnetic field that is described by the Fourier coefficients $X_1, X_2, \cdots$ or directly in terms of the magnetic field configuration itself by

$$\psi(B(x, y, z)) = Pexp\left(-\int\int \frac{\mathbf{B(x_1)} \cdot \mathbf{B(x_2)}}{16\pi^3 \hbar c r_{12}^2} d^3x_1 d^3x_2\right).$$

P being a normalization factor. Let us consider a configuration where the magnetic field is everywhere zero except in a region of dimension l, where it is of the order of $\sim \Delta B$. The probability amplitude for this configuration would be proportional to

$$\exp\left[-\left((\Delta B)^2 l^4/\hbar c\right)\right]$$

So the energy of fluctuation in a volume of length l is given by finally [45, 197, 198]

$$B^2 \sim \frac{\hbar c}{l} \tag{3.1}$$

We will return to (3.1) subsequently but observe that if in (3.1) above l is taken to be the Compton wavelength of a typical elementary particle, then we recover its energy mc^2, as can be easily verified. In the previous Chapter, we had seen how inertial mass and energy can be deduced on the basis of viscous resistance to the ZPF. We will also deduce this from Quantum Mechanical effects within the Compton scale. The above gives us back this result in the context of the ZPF. In any case (3.1) shows the inverse dependence of the length scale and the energy (or momentum).
It may be reiterated that Einstein himself had believed that the electron was a result of such a condensation from the background electromagnetic field (Cf.[199, 130] for details). The above result is very much in this spirit. In the sequel we also take the pion to represent a typical elementary particle, as in the literature.

To proceed, as there are $N \sim 10^{80}$ such particles in the Universe, we get, consistently,

$$Nm = M \tag{3.2}$$

where M is the mass of the Universe. It must be remembered that the energy of gravitational interaction between the particles is very much insignificant compared to the above electromagnetic considerations.
In the following we will use N as the sole cosmological parameter.
We next invoke the well known relation [200, 131, 126]

$$R \approx \frac{GM}{c^2} \tag{3.3}$$

where M can be obtained from (3.2). We can arrive at (3.3) in different ways. For example, in a uniformly expanding Friedman Universe, we have

$$\dot{R}^2 = 8\pi G\rho R^2/3$$

In the above if we substitute $\dot{R} = c$ at R, the radius of the universe, we get (3.3). Another proof will be given later in Section 3.10.
We now use the fact that given N particles, the (Gaussian)fluctuation in the particle number is of the order $\sqrt{N}$[126, 125, 175, 176, 173, 174], while a typical time interval for the fluctuations is $\sim \hbar/mc^2$, the Compton time, the fuzzy interval we encountered in the previous Chapter within which there is no meaningful physics. We will come back to this point later in this Chapter, in the context of the minimum Planck scale. So particles are created and destroyed−−but the ultimate result is that $\sqrt{N}$ particles are created just as this is the nett displacement in a random walk of unit step. So we have,

$$\frac{dN}{dt} = \frac{\sqrt{N}}{\tau} \tag{3.4}$$

whence on integration we get, (remembering that we are almost in the continuum region that is, $\tau \sim 10^{-23} sec \approx 0$),

$$T = \frac{\hbar}{mc^2}\sqrt{N} \tag{3.5}$$

We can easily verify that the equation (3.5) is indeed satisfied where T is the age of the Universe. Next by differentiating (3.3) with respect to t we get

$$\frac{dR}{dt} \approx HR \tag{3.6}$$

where H in (3.6) can be identified with the Hubble Constant, and using (3.3) is given by,

$$H = \frac{Gm^3c}{\hbar^2} \tag{3.7}$$

Equations (3.2), (3.3) and (3.5) show that in this formulation, the correct mass, radius, Hubble constant and age of the Universe can be deduced given N, the number of particles, as the sole cosmological or large scale parameter. We observe that at this stage we are not invoking any particular dynamics−−the expansion is due to the random creation of particles from the ZPF background. Equation (3.7) can be written as

$$m \approx \left(\frac{H\hbar^2}{Gc}\right)^{\frac{1}{3}} \tag{3.8}$$

Equation (3.8) has been empirically known as an "accidental" or "mysterious" relation. As observed by Weinberg [17], this is unexplained: it relates a single cosmological parameter H to constants from microphysics. We will touch upon this micro-macro nexus again. In our formulation, equation (3.8) is no longer a mysterious coincidence but rather a consequence of the theory.

As (3.7) and (3.6) are not exact equations but rather, order of magnitude relations, it follows, on differentiating (3.6) that a small cosmological constant $\wedge$ is allowed such that

$$\wedge \leq 0(H^2)$$

This is consistent with observation and shows that $\wedge$ is very small−−this has been a puzzle, the so called cosmological constant problem alluded to, because in conventional theory, it turns out to be huge [188]. But it poses no problem in this formulation. This is because of the characterization of the ZPF as independent and primary in our formulation this being the mysterious dark energy. We shall further characterize $\wedge$ later in this Chapter.

To proceed we observe that because of the fluctuation of $\sim \sqrt{N}$ (due to the ZPF), there is an excess electrical potential energy of the electron, which in fact we identify as its inertial energy. That is [175, 126],

$$\sqrt{N}e^2/R \approx mc^2.$$

On using (3.3) in the above, we recover the well known Gravitation-Electromagnetism ratio viz.,

$$e^2/Gm^2 \sim \sqrt{N} \approx 10^{40} \tag{3.9}$$

or without using (3.3), we get, instead, the well known so called Weyl-Eddington formula,

$$R = \sqrt{N} l \tag{3.10}$$

(It appears that (3.10) was first noticed by H. Weyl [110]). Infact (3.10) is the spatial counterpart of (3.5). If we combine (3.10) and (3.3), we get,

$$\frac{Gm}{lc^2} = \frac{1}{\sqrt{N}} \propto T^{-1} \tag{3.11}$$

where in (3.11), we have used (3.5). Following Dirac (cf.also [201]) we treat G as the variable, rather than the quantities m, l, c and $\hbar$ which we will call micro physical constants because of their central role in atomic (and sub atomic) physics.
Next if we use G from (3.11) in (3.7), we can see that

$$H = \frac{c}{l} \quad \frac{1}{\sqrt{N}} \tag{3.12}$$

Thus apart from the fact that H has the same inverse time dependance on T as G, (3.12) shows that given the microphysical constants, and N, we can deduce the Hubble Constant also, as from (3.12) or (3.7).
Using (3.2) and (3.3), we can now deduce that

$$\rho \approx \frac{m}{l^3} \quad \frac{1}{\sqrt{N}} \tag{3.13}$$

Next (3.10) and (3.5) give,

$$R = cT \tag{3.14}$$

Equations (3.13) and (3.14) are consistent with observation.
Finally, we observe that using $M, G \text{and} H$ from the above, we get

$$M = \frac{c^3}{GH}$$

This relation is required in the Friedman model of the expanding Universe (and the Steady State model too). In fact if we use in this relation, the expression,

$$H = c/R$$

which follows from (3.12) and (3.10), then we recover (3.3). We will be repeatedly using these relations in the sequel.
As we saw the above model predicts a dark energy driven ever expanding and accelerating Universe with a small cosmological constant while the density keeps decreasing. Moreover mysterious large number relations like

(3.7), (3.13) or (3.10) which were considered to be miraculous accidents now follow from the underlying theory. This seemed to go against the accepted idea that the density of the Universe equalled the critical density required for closure and that aided by dark matter, the Universe was decelerating. However, as noted, from 1998 onwards, following the work of Perlmutter, Schmidt and co-workers, these otherwise apparently heretic conclusions have been vindicated.

It may be mentioned that the observational evidence for an accelerating Universe was the American Association for Advancement of Science's Breakthrough of the Year, 1998 while the evidence for nearly seventy five percent of the Universe being Dark Energy, based on the Wilkinson Microwave Anisotropy Probe (WMAP) and the Sloan Sky Digital Survey was the Breakthrough of the Year, 2003 [202, 171].

3.3 Issues and Ramifications

Cosmologies with time varying G have been considered in the past, for example in the Brans-Dicke theory or in the Dirac large number theory or by Hoyle [203–207]. In the case of the Dirac cosmology, the motivation was Dirac's observation that the supposedly large number coincidences involving N, the number of elementary particles in the universe had an underlying message if it is recognized that

$$\sqrt{N} \propto T \tag{3.15}$$

where T is the age of the universe. Equation (3.15) which is essentially equation (3.5) lead to a G decreasing inversely with time.

The Brans-Dicke cosmology arose from the work of Jordan who was motivated by Dirac's ideas to try and modify General Relativity suitably. In this scheme the variation of G could be obtained from a scalar field ϕ which would satisfy a conservation law. This scalar tensor gravity theory was further developed by Brans and Dicke, in which G was inversely proportional to the variable field ϕ. (It may be mentioned that more recently the ideas of Brans and Dicke have been further generalized.)

In the Hoyle-Narlikar steady state model, it was assumed that in the Machian sense the inertia of a particle originates from the rest of the matter present in the universe. This again leads to a variable G. The above references give further details of these various schemes and their shortcomings which have lead to their falling out of favour.

Then there is the author's cosmology seen briefly in the last section in which particles are fluctuationally created from a background dark energy, in an inflationary type phase transition as we will briefly see, and this leads to a scenario of an accelerating universe with a small cosmological constant. Moreover, to reiterate, in the author's cosmology the various supposedly miraculous large number coincidences as also the otherwise inexplicable Weinberg formula which gives the mass of an elementary particle in terms of the gravitational constant and the Hubble constant are also deduced from the underlying theory rather than being ad hoc as seen in the previous Section. We will discuss this point in Section 3.8. The gravitational constant is given from (3.11) by

$$G = \frac{G_0}{T} \tag{3.16}$$

where T is time (the age of the universe) and G_0 is a constant. Furthermore, other routine effects like the precession of the perihelion of Mercury and the bending of light, the flattening of rotational curves of galaxies and so on are also explained in this model as will be discussed below. Moreover in this model, Λ is given by $\Lambda \leq 0(H^2)$ and shows the inverse dependence $1/T^2$ on time. We will also see that there is observational evidence for (3.16).

With this background, we now give some tests for equation (3.16).

3.4 Tests

There have been some observations, like the precession of the perihelion of Mercury or the bending of light, which could not be explained by Newtonian mechanics. As is well known, it was a triumph of General Relativity, that these could be accounted for. The question arises, if the new theory, particularly (3.16), could also explain these phenomena. We first deduce using (3.16), the perihelion precession of Mercury [208]. We observe that from (3.16) it follows that [43]

$$G = G_o(1 - \frac{t}{t_o}) \tag{3.17}$$

where G_o is the present value of G and t_o is the present age of the Universe and t the (relatively small) time elapsed from the present epoch. Similarly one could deduce that (cf.ref.[43]),

$$r = r_o \left(\frac{t_o}{t_o + t}\right) \tag{3.18}$$

We next use Kepler's Third law[209]:

$$\tau = \frac{2\pi a^{3/2}}{\sqrt{GM}} \tag{3.19}$$

τ is the period of revolution, a is the orbit's semi major axis, and M is the mass of the sun. Denoting the average angular velocity of the planet by

$$\dot{\Theta} \equiv \frac{2\pi}{\tau},$$

it follows from (3.17), (3.18) and (3.19) that

$$\dot{\Theta} - \dot{\Theta}_o = \dot{\Theta}_0 \frac{t}{t_o},$$

where the subscript o refers to the present epoch.
Whence,

$$\omega(t) \equiv \Theta - \Theta_o = \frac{\pi}{\tau_o t_o} t^2 \tag{3.20}$$

Equation (3.20) gives the average perihelion precession at time "t". Specializing to the case of Mercury, where $\tau_o = 0.25$ year, it follows from (3.20) that the average precession per year at time "t" is given by

$$\omega(t) = \frac{4\pi t^2}{t_0} \tag{3.21}$$

Whence, considering $\omega(t)$ for years $t = 1, 2, \cdots, 100$, we can obtain from (3.21), the correct perihelion precession per century as [208],

$$\omega = \sum_{n=1}^{100} \omega(n) \approx 43'',$$

if the age of the universe is taken to be $\approx 2 \times 10^{10}$ years.
Conversely, if we use the observed value of the precession in (3.21), we can get back the above age of the universe.
Interestingly it can be seen from (3.21), that the precession depends on the epoch.
We next demonstrate that orbiting objects will have an anamolous inward radial acceleration.
Using the well known equation for Keplarian orbits (cf.ref.[209]),

$$\frac{1}{r} = \frac{GMm^2}{l^2}(1 + ecos\Theta) \tag{3.22}$$

$$\dot{r}^2 \approx \frac{GM}{r} - \frac{l^2}{m^2 r^2} \tag{3.23}$$

l being the orbital angular momentum constant and e the eccentricity of the orbit, we can deduce such an extra inward radial acceleration, on differentiation of (3.23) and using (3.17) and (3.18),

$$a_r = \frac{GM}{2t_o r\dot{r}} \tag{3.24}$$

It can be easily shown from (3.22) that (on the average),

$$\dot{r} \approx \frac{eGM}{rv} \tag{3.25}$$

For a nearly circular orbit $rv^2 \approx GM$, whence use of (3.25) in (3.24) gives,

$$a_r \approx v/2t_o e \tag{3.26}$$

For the earth, (3.26) gives an anomalous inward radial acceleration $\sim 10^{-9} cm/sec^2$, which is known to be the case [210].

We could also deduce a progressive decrease in the eccentricity of orbits. Indeed, e in (3.22) is given by

$$e^2 = 1 + \frac{2El^2}{G^2m^3M^2} \equiv 1 + \gamma, \gamma < 0.$$

Use of (3.17) in the above and differenciation, leads to,

$$\dot{e} = \frac{\gamma}{et_o} \approx -\frac{1}{et_o} \approx -\frac{10^{-10}}{e} \text{per year},$$

if the orbit is nearly circular. (Variations of eccentricity in the usual theory have been extensively studied (cf.ref.[211] for a review).) On the other hand, for open orbits, $\gamma > 0$, the eccentricity would progressively increase.

We finally consider the anomalous accelerations given in (3.24) and (3.26) in the context of space crafts leaving the solar system.

If in (3.24) we use the fact that $\dot{r} \leq v$ and approximate

$$v \approx \sqrt{\frac{GM}{r}},$$

we get,

$$a_r \geq \frac{1}{et_o}\sqrt{\frac{GM}{r}}$$

For $r \sim 10^{14} cm$, as is the case of the space crafts Pioneer 10 or Pioneer 11, this gives, $a_r \geq 10^{-11} cm/sec^2$ We will soon further refine this result.

Interestingly Anderson et al.,[212] claim to have observed an anomalous inward acceleration of $\sim 10^{-8} cm/sec^2$ for more than a decade.

3.5 Other Consequences

We could also explain the correct gravitational bending of light. Infact in Newtonian theory too we obtain the bending of light, though the amount is half that predicted by General Relativity[43, 213–215]. In the Newtonian theory we can obtain the bending from the well known orbital equations (Cf.also(3.22)),

$$\frac{1}{r} = \frac{GM}{L^2}(1 + ecos\Theta) \tag{3.27}$$

where M is the mass of the central object, L is the angular momentum per unit mass, which in our case is bc, b being the impact parameter or minimum approach distance of light to the object, and e the eccentricity of the trajectory is given by

$$e^2 = 1 + \frac{c^2L^2}{G^2M^2} \tag{3.28}$$

For the deflection of light α, if we substitute $r = \pm\infty$, and then use (3.28) we get

$$\alpha = \frac{2GM}{bc^2} \tag{3.29}$$

This is half the General Relativistic value.
We now note that the effect of time variation of r is given by equation (3.18)(cf.ref.[208]). Using (3.18) the well known equation for the trajectory is given by,

$$u" + u = \frac{GM}{L^2} + u\frac{t}{t_0} + 0\left(\frac{t}{t_0}\right)^2 \tag{3.30}$$

where $u = \frac{1}{r}$ and primes denote differenciation with respect to Θ.
The first term on the right hand side represents the Newtonian contribution while the remaining terms are the contributions due to (3.18). The solution of (3.30) is given by

$$u = \frac{GM}{L^2}\left[1 + ecos\left\{\left(1 - \frac{t}{2t_0}\right)\Theta + \omega\right\}\right] \tag{3.31}$$

where ω is a constant of integration. Corresponding to $-\infty < r < \infty$ in the Newtonian case we have in the present case, $-t_0 < t < t_0$, where t_0 is large and infinite for practical purposes. Accordingly the analogue of the reception of light for the observer, viz., $r = +\infty$ in the Newtonian case is obtained by taking $t = t_0$ in (3.31) which gives

$$u = \frac{GM}{L^2} + ecos\left(\frac{\Theta}{2} + \omega\right) \tag{3.32}$$

Comparison of (3.32) with the Newtonian solution obtained by neglecting terms $\sim t/t_0$ in equations (3.18),(3.30) and (3.31) shows that the Newtonian Θ is replaced by $\frac{\Theta}{2}$, whence the deflection obtained by equating the left side of (3.32) to zero, is

$$cos\Theta\left(1-\frac{t}{2t_0}\right)=-\frac{1}{e} \tag{3.33}$$

where e is given by (3.28). The value of the deflection from (3.33) is twice the Newtonian deflection given by (3.29). That is the deflection α is now given not by (3.29) but by the formula,

$$\alpha=\frac{4GM}{bc^2}, \tag{3.34}$$

The relation (3.34) is the correct observed value and is the same as the General Relativistic formula which however is obtained by a different route [215, 39, 216].

We now come to the problem of galactic rotational curves mentioned earlier (cf.ref.[43]). We would expect, on the basis of straightforward dynamics that the rotational velocities at the edges of galaxies would fall off according to

$$v^2\approx\frac{GM}{r} \tag{3.35}$$

However it is found that the velocities tend to a constant value,

$$v\sim 300km/sec \tag{3.36}$$

This, as noted, has lead to the postulation of the as yet undetected additional matter alluded to, the so called dark matter.(However for an alternative view point Cf.[217]). We observe that from (3.18) it can be easily deduced that[130, 218]

$$a\equiv(\ddot{r}_o-\ddot{r})\approx\frac{1}{t_o}(t\ddot{r}_o+2\dot{r}_o)\approx-2\frac{r_o}{t_o^2} \tag{3.37}$$

as we are considering infinitesimal intervals t and nearly circular orbits. Equation (3.37) shows (Cf.ref[208] also) that there is an anomalous inward acceleration, as if there is an extra attractive force, or an additional central mass, as indeed we saw a little earlier.

So,

$$\frac{GMm}{r^2}+\frac{2mr}{t_o^2}\approx\frac{mv^2}{r} \tag{3.38}$$

From (3.38) it follows that

$$v \approx \left(\frac{2r^2}{t_o^2} + \frac{GM}{r} \right)^{1/2} \tag{3.39}$$

From (3.39) it is easily seen that at distances within the edge of a typical galaxy, that is $r < 10^{23} cms$ the equation (3.35) holds but as we reach the edge and beyond, that is for $r \geq 10^{24} cms$ we have $v \sim 10^7 cms$ per second, in agreement with (3.36). In fact as can be seen from (3.39), the first term in the square root has an extra contribution (due to the varying G) which is roughly some three to four times the second term, as if there is an extra mass, roughly that much more.

Thus the time variation of G explains observation without invoking dark matter. There could be other explanations, too. The author and A.D. Popova have argued that if the three dimensionality of space asymptotically falls off, then the above can be explained [219]. Yet another prescription was given by Milgrom [220] who approached the problem by modifying Newtonian dynamics at large distances. This approach is purely phenomenological.

The idea was that perhaps standard Newtonian dynamics works at the scale of the solar system but at galactic scales involving much larger distances perhaps the situation is different. However a simple modification of the distance dependence in the gravitation law, as pointed by Milgrom would not do, even if it produced the asymptotically flat rotation curves of galaxies. Such a law would predict the wrong form of the mass velocity relation. So Milgrom suggested the following modification to Newtonian dynamics: A test particle at a distance r from a large mass M is subject to the acceleration a given by

$$a^2/a_0 = MGr^{-2}, \tag{3.40}$$

where a_0 is an acceleration such that standard Newtonian dynamics is a good approximation only for accelerations much larger than a_0. The above equation however would be true when a is much less than a_0. Both the statements in (3.40) can be combined in the heuristic relation

$$\mu(a/a_0)a = MGr^{-2} \tag{3.41}$$

In (3.41) $\mu(x) \approx 1$ when $x >> 1$, and $\mu(x) \approx x$ when $x << 1$. It must be stressed that (3.40) or (3.41) are not deduced from any theory, but rather are an ad hoc prescription to explain observations. Interestingly it must be mentioned that most of the implications of Modified Newtonian Dynamics

or MOND do not depend strongly on the exact form of μ.
It can then be shown that the problem of galactic velocities is now solved [221–225].

3.6 The Anomalous Acceleration of the Pioneer Spacecrafts

The inexplicable anomalous accelerations of the Pioneer spacecrafts already alluded to, which have been observed by J.D. Anderson and coworkers at the Jet Propulsion Laboratory for well over a decade [226, 212] have posed a puzzle. This can be explained by (3.16), in a simple way as follows: In fact from the usual orbital equations we have [227]

$$v\dot{v} \approx -\frac{GM}{2tr}(1 + ecos\Theta) - \frac{GM}{r^2}\dot{r}(1 + ecos\Theta)$$

v being the velocity of the spacecraft and t is the time in general. It must be observed that the first term on the right side is the new effect due to (3.16). There is now an anomalous acceleration given by

$$a_r = \langle \dot{v} \rangle \text{anom} = \frac{-GM}{2trv}(1 + ecos\Theta)$$

$$\approx -\frac{GM}{2t\lambda}(1 + e)^3$$

where

$$\lambda = r^4\dot{\Theta}^2$$

If we insert the values for the Pioneer spacecrafts we get

$$a_r \sim -10^{-7} cm/sec^2$$

This is the anomalous acceleration reported by Anderson and co-workers. We will next deduce that the equation (3.16) explains correctly the observed decrease in the orbital period of the binary pulsar $PSR\,1913 + 16$, which has also been attributed to as yet undetected gravitational waves [228].

3.7 The Binary Pulsar

It may be observed that the energy E of two masses M and m in gravitational interaction at a distance L is given by

$$E = \frac{GMm}{L} = \text{constant} \tag{3.42}$$

We note that if this energy decreases by any mechanism, for example by the emission of gravitational waves, or by the decrease of G, then because of (3.42), there is a compensation by the decrease in the orbital length and orbital period. This is the standard General Relativistic explanation for the binary pulsar $PSR\,1913 + 16$. We will show that the same holds good, if we are given instead, (3.16). That is, we will not invoke gravitational waves. In this case we have, from (3.42)

$$\frac{\mu}{L} \equiv \frac{GMm}{L} = \text{const.} \tag{3.43}$$

Using (3.16) we can write, for a time increase t,

$$\mu = \mu_0 - Kt \tag{3.44}$$

where we have

$$K \equiv \dot{\mu} \tag{3.45}$$

In (3.45) $\dot{\mu}$ can be taken to be a constant in view of the fact that G varies very slowly with time as can be seen from (3.16). Specifically we have

$$G(T+t) = G(T) - t\frac{G}{T} + \frac{t^2}{2}\frac{G}{T^2} + \cdots \approx G(T) - t\frac{G}{T} \tag{3.46}$$

where T is the age of the universe and t is an incremental time. Whence using (3.46), K in (3.45) is given by

$$K \propto \frac{G}{T}$$

and so

$$\dot{K} \sim \frac{G}{T^2} \approx 0$$

So (3.43) requires

$$L = L_0(1 - \alpha K)$$

Whence on using (3.44) we get

$$\alpha = \frac{t}{\mu_0} \tag{3.47}$$

Let us now consider t to be the period of revolution in the case of the binary pulsar. Using (3.47) it follows that

$$\delta L = -\frac{L_0 t K}{\mu_0} \tag{3.48}$$

We also know (Cf.ref.[227])

$$t = \frac{2\pi}{h}L^2 = \frac{2\pi}{\sqrt{\mu}} \tag{3.49}$$

$$t^2 = \frac{4\pi^2 L^3}{\mu}, \tag{3.50}$$

h being the usual unit angular momentum and μ has the units $gm\, cm^4 sec^{-1}$. Using (3.48), (3.49) and (3.50), a little manipulation gives

$$\delta t = -\frac{2t^2 K}{\mu_0} \tag{3.51}$$

(3.48) and (3.51) show that there is a decrease in the size of the orbit, as also in the orbital period. Such a decrease in the orbital period has been observed in the case of binary pulsars in general [228, 229].

Let us now apply the above considerations to the specific case of the binary pulsar $PSR\,1913+16$ observed by Taylor and coworkers (Cf.ref.[229]). In this case it is known that, t is 8 hours while v, the orbital speed is $3\times 10^7 cms$ per second. It is easy to calculate from the above

$$\mu_0 = 10^4 \times v^3 \sim 10^{26}$$

which gives $M \sim 10^{33} gms$, which of course agrees with observation. Further we get using (3.15) and (3.44)

$$\Delta t = \eta \times 10^{-5} sec/yr, \eta <\approx 8 \tag{3.52}$$

Indeed (3.52) is in good agreement with the carefully observed value of $\eta \approx 7.5$ (Cf.refs.[228, 229]).

It should also be remarked that in the case of gravitational radiation, there are some objections relevant to the calculation (Cf.ref.[228]).

Finally, we may point out that a similar shrinking in size with time can be expected of galaxies themselves, and in general, gravitationally bound systems. We will see a special case for the solar system in the next Section.

3.8 Change in Orbital Parameters

To consider the above result in a more general context, we come back to the well known orbital equation [227]

$$d^2u/d\Theta^2 + u = \mu_0/h^2 \tag{3.53}$$

where $\mu_0 = GM$ and u is the usual inverse of radial distance.
M is the mass of the central object and $h = r^2 d\Theta/dt$—–a constant. The solution of (3.53) is well known,

$$lu = 1 + ecos\Theta$$

where $l = h^2/\mu_0$.
It must be mentioned that in the above purely classical analysis, there is no precession of the perihelion.

We now replace μ_0 by μ and also assume μ to be varying slowly because G itself varies slowly and uniformly, as noted earlier:

$$\dot{\mu} = d\mu/dt = K, \text{a} \quad \text{constant} \tag{3.54}$$

remembering that $\dot{K} \sim 0(1/T^2)$ and so can be neglected.

Using (3.54) in (3.53) and solving the orbital equation (3.53), the solution can now be obtained as

$$u = 1/l + (e/l)cos\Theta + Kl^2\Theta/h^3 + Kl^2e\Theta cos\Theta/h^3 \tag{3.55}$$

Keeping terms up to the power of 'e' and $(K/\mu_0)^2$, the time period 'τ' for one revolution is given to this order of approximation by

$$\tau = 2\pi L^2/h \tag{3.56}$$

From (3.55)

$$L = l - \frac{Kl^4\Theta}{h^3} \tag{3.57}$$

Substituting (3.57) in (3.56) we have

$$\tau = \frac{2\pi}{h}\left(l^2 - \frac{2Kl^5\Theta}{h^3}\right) \tag{3.58}$$

The second term in (3.58) represents the change in time period for one revolution. The decrease of time period is given by

$$\delta\tau = 8\pi^2 l^3 K/\mu_0^2 \tag{3.59}$$

The second term in (3.57) indicates the decrease in latus-rectum.

For one revolution the change of latus-rectum is given by

$$\delta l = 2\pi K l^{2.5}/\mu_0^{1.5} \tag{3.60}$$

In the solar system, we have,

$$K = 898800\, cm\, gm$$

Using K and μ_0 to find the change in time period and the latus rectum in the varying G case by substituting in (3.59) and (3.60) respectively for Mercury we get

$$\delta T = 1.37 \times 10^{-5} sec/rev$$

$$\delta l = 4.54 cm/rev \tag{3.61}$$

We observe that the equations (3.59), (3.60) or (3.61) show a decrease in distance and in the time of revolution. If we use for the planetary motion, the General Relativistic analogue of (3.53), viz.,

$$\frac{d^2u}{d\Theta^2} + u = \frac{\mu_0}{h^2}(1 + 3h^2u^2),$$

then while we recover the precession of the perihelion of Mercury, for example, there is no effect similar to (3.59), (3.60) or (3.61). On the other hand this effect is very minute−−just a few centimeters per year in the case of the earth−−and only protracted careful observations can detect it. Moreover these changes could also be masked at least partly, by gravitational and other perturbations.

However as noted, the decrease of the period in (3.59) has been observed in the case of Binary Pulsars.

3.9 Remarks

i) With regard to the time variation of G, it must be mentioned that without reference to the tests alluded to, different observations have yielded different values. Observations on the earth, in the solar system and with Pulsars have yielded for $\frac{\dot{G}}{G}$ a value $\sim 10^{-10}/yr$ as in (3.16). However other model dependent observations have yielded values $\sim 10^{-11}/yr$ and $10^{-12}/yr$ [230].
ii) We may also remark that Fred Hoyle had suggested that a variation in G could be responsible for the Tectonic activity on the Earth in the following way [231]. As the gravitational effect weakens, the Earth tends to expand, leading to a cracking of its crust. Perhaps this could explain the formation of continents. Further, the internal pressures of the fluid layers below the crust may be strong enough to move the broken crust or continents, thus leading to the continental drift.
iii) There has been a wealth of data from the WMAP (Cf. for example [232, 233]). One of the intriguing findings is that the dark energy domination and the CMB power suppression, both occur around the same red

shift and energy scale−−corresponding to the Hubble radius $\sim 10^{-33}eV$. As has been earlier pointed out [21] this raises three disturbing questions, viz., the small magnitude, the so called tuning problem and why this should occur during this epoch, that is the coincidence problem, and finally why do both coincidences occur at the energy scale of our present Hubble radius of $10^{-33}eV$. The question has also been asked, "Does all this suggest new physics?".

We now show that this is explained in terms of the Planck scale underpinning for the universe, as has been discussed in detail by the author (Cf. refs.[157, 158, 234, 175] and references therein) and indeed is an observational confirmation for the cosmological model discussed earlier. We summarize the main results. The universe as we will see later, can be considered to have an underpinning of $\bar{N}$ Planck oscillators. Further we have

$$R = \sqrt{\bar{N}} l_P \tag{3.62}$$

In (3.62) $R(\sim 10^{-33}eV)$ is the radius of the universe and $l_P \sim 10^{-33}cms$ is the Planck length. It can also be shown that $\bar{N} \sim 10^{120}$. All this is similar to the earlier Compton scale considerations. Moreover there is a minimum mass (or minimum dark energy scale) in the universe which is given by

$$m = m_P/\sqrt{\bar{N}} \sim 10^{-65}gm \sim 10^{-33}eV \tag{3.63}$$

It can now be seen that the puzzling small energy scale $\sim 10^{-33}eV$ referred to earlier is exactly the same as the minimum allowable energy (or mass m) given in (3.63). Moreover the Compton wavelength of this mass is exactly $\sim R$, the Hubble radius. What all this means is, at this point of time in the universe, there is a minimum energy of the background dark energy $\sim 10^{-33}eV$.

Another way of deriving this result is directly from (3.1). If for the extension l, we take the radius of the universe itself, then we get the residual energy $10^{-33}eV$. We will return to this result again, later.

iv) We note that there is an acceleration $\sim 10^{-7}cm/sec$ at R. But this is true everywhere as all points of space are on the same footing. This is the so called empirical "Milgrom" acceleration.

The Milgrom acceleration can be written as

$$a_0 \sim H(\sim 10^{-7}cmsec^{-2})(\sim c^2/R) \tag{3.64}$$

where H is the Hubble constant. In fact this follows from the earlier varying G theory. For, we have in this case from (3.37),

$$a_0 \sim r/t_0^2$$

Feeding the values of r, the radius of the universe $= ct_0$ and the fact that $H \sim \frac{1}{t_0}$, we get (3.64), which now shows up no longer as an ad hoc prescription but rather as a consequence of the theory. Curiously enough, a particle obeying Newtonian dynamics, which has the acceleration (3.64) over the life time of the universe, attains the velocity of light and moreover covers a distance equalling the size of the universe.
It may be noted that the Boomerang results are in tune with MOND rather than the Dark Matter scenario, the WMAP model notwithstanding [235].
We reiterate that the variation of G discussed in Sections 5 and 6, shows that there is an inward acceleration in gravitationally bound systems−−this would imply as noted that such systems (galaxies included) would tend to become progressively smaller, as with binary pulsar orbits−−in the absence of other dynamical considerations.
v) We have alluded to relations like (3.5), (3.7), (3.8), (3.9) and (3.10), the so called Large Number relations. In all these cases it turns out that T, the age of the Universe is proportional to a suitable power of N the number of particles in the Universe. Rather than dismiss these relations as mere coincidences, Dirac as noted suspected that these pointed to a relationship with time. In his words [236]:
"I call this principle the

Large Numbers Hypothesis

According to it, all the very large dimensionless numbers, which turn up in Nature, are related to one another, just like $t = 7 \times 10^{39}$ and e^2/Gm_em_p.
"There is one further very large dimensionless number which we have to take into consideration. That is the total mass of the universe when expressed in units of, say, the proton mass. That will be, if you like, the total number of protons and neutrons in the Universe. It may be, of course, that the Universe is infinite and that, therefore, this total number is infinite. In that case we should not be able to talk about it. Yet we can use another number to replace it. We need only consider that portion of the Universe which is sufficiently close to us for the velocity of recession to be less than, let us say, half the velocity of light. We are then considering just a certain chunk of this infinite Universe, for which recession velocities are less than half the velocity of light. We then ask, what is the total mass of this chunk of the Universe? That again will be a very Large Number and will replace the total mass of the Universe, to give us a definite number when the Universe is infinite.
"We may try to estimate this total mass using the mass of those stellar

objects which we can observe, and making an allowance for unobservable matter. We do not know very well how big that allowance should be: there may be quite a lot of unobservable matter in the form of intergalactic gas or black holes or things like that. Still, it is probable that the amount of dark matter is not very much greater than the amount of visible matter. If you make an assumption of that kind, you find that the total mass, in terms of the proton mass, is

$$\frac{\text{total mass}}{\text{proton mass}} = 10^{78},$$

with a suitable factor allowed for the invisible matter. We, therefore get a number which is, roughly, the square of t (in atomic units).

"Now, according to the Large Number Hypothesis, all these very large dimensionless numbers should be connected together. We should then expect that

$$\frac{\text{total mass}}{\text{proton mass}} = 10^{78} :: t^2,$$

Using the same argument again, we are therefore led to think that the total number of protons in the Universe is increasing proportionally to t^2. Thus, there must be creation of matter in the Universe, a continuous creation of matter.

"There have been quite a number of cosmological theories working with continuous creation of matter. A theory like that was very much developed by Hoyle and others. The continuous creation which I am proposing here is entirely different from that. Their continuous creation theory was introduced as a rival to the Big Bang theory, and it is not in favor at the present time.

"The continuous creation which I have here is essentially different from Hoyle's continuous creation, because Hoyle was proposing a steady state of the Universe, with continuous creation to make up for the matter which is moving beyond our region of vision by the expansion. In his steady-state theory, he had G constant. Now, in the present theory, G is varying with time, and that makes an essential difference.

"I propose a theory where there is continuous creation of matter, together with this variation of G. Both the assumption of continuous creation and the variation of G follow from the Large Numbers Hypothesis.

This continuous creation of matter must be looked upon as something quite independent of known physical processes. According to the ordinary physical processes, which we study in the laboratory, matter is conserved. Here we have direct nonconservation of matter. It is, if you like, a new kind of radioactive process for which there is nonconservation of matter and by which particles are created where they did not previously exist. The effect is very small, because the number of particles created will be appreciable only we wait for a very long time interval compared with the age of the Universe."

There were however, some inconsistencies in this Dirac cosmology. For instance, if there would be no particle creation, then we would have,

$$R \propto T^{1/3}$$

He vacillated over the decades between versions using the conservation of energy and also violating it.

In our cosmology, using fluctuations, all these apparently disparate relations are derived from underlying principles, not to mention the prediction of a dark energy driven accelerating Universe with a permissible cosmological constant. That is what science is all about - finding a minimum set of principles to explain a maximal set of observations.

vi) The above cosmology as we saw exhibits a time variation of the gravitational constant of the form of (3.16),

$$G = \frac{\beta}{T} \tag{3.65}$$

Indeed as noted, this is true in a few other schemes also, including the so called Brans-Dicke and Dirac cosmologies. We have also shown that such a time variation can explain the precession of the perihelion of Mercury and several other effects (Cf. [208]). For example it can also provide an alternative explanation for dark matter and the bending of light (while the Cosmic Microwave Background Radiation is also explained (Cf.[130])) and so on.

It is also possible to deduce the "existence" of gravitational waves given (3.65) (or (3.16)). This is suggested by the considerations of Section 3.7. To see this quickly let us consider the Poisson equation for the metric $g_{\mu\nu}$

$$\nabla^2 g_{\mu\nu} = G\rho u_\mu u_\nu \tag{3.66}$$

The solution of (3.66) is given by

$$g_{\mu\nu} = G \int \frac{\rho u_\mu u_\nu}{|\vec{r} - \vec{r'}|} d^3\vec{r} \tag{3.67}$$

Indeed equations similar to (3.66) and (3.67) hold for the Newtonian gravitational potential also. If we use the second time derivative of G from (3.65) in (3.67), along with (3.66), we can immediately obtain the D'alembertian wave equation for gravitational waves, instead of the Poisson equation:

$$Dg_{\mu\nu} \approx 0$$

vii) Recently a small variation with time of the fine structure constant has been detected and reconfirmed by Webb and coworkers [237, 238]. In a sense, this is shocking−−the fine structure constant has been a hallowed quantity and that it should vary with time is not easy to accept. However, this observation is consistent with the above cosmology. We can see this as follows. We use an equation due to Kuhne [239]

$$\frac{\dot{\alpha}_z}{\alpha_z} = \alpha_z \frac{\dot{H}_z}{H_z}, \tag{3.68}$$

If we now use the fact that the cosmological constant Λ is given by

$$\Lambda \leq 0(H^2) \tag{3.69}$$

as can be seen from (3.6), in (3.68), we get using (3.69),

$$\frac{\dot{\alpha}_z}{\alpha_z} = \beta H_z \tag{3.70}$$

where $\beta < -\alpha_z < -10^{-2}$.
Equation (3.70) can be shown to be the same as

$$\frac{\dot{\alpha}_z}{\alpha_z} \approx -1 \times 10^{-5} H_z. \tag{3.71}$$

which is the same as Webb's result.
We give another derivation of (3.71) in the above context wherein, as the number of particles in the Universe increases with time, we go from the Planck scale to the Compton scale (a theme to which we will return soon). This can be seen as follows: In equation (3.9), if the number of particles in the Universe, $N = 1$, then the mass m would be the Planck mass. In this case the classical Schwarzschild radius of the Planck mass would equal its Quantum Mechanical Compton wavelength. To put it another way, all the energy would be gravitational (Cf.[130] for details). However as the number of particles N increases with time, according to (3.5), Gravitation and Electromagnetism get differentiated and we get (3.9) and the Compton scale.
It is known that the Compton length, due to zitterbewegung causes a correction to the electrostatic potential which an orbiting electron experiences, rather like the Darwin term [181].
Infact we have

$$\langle \delta V \rangle = \langle V(\vec{r} + \delta\vec{r}) \rangle - V\langle(\vec{r})\rangle$$

$$= \langle \delta r \frac{\partial V}{\partial r} + \frac{1}{2} \sum_{ij} \delta r_i \delta r_j \frac{\partial^2 V}{\partial r_i \partial r_j} \rangle$$

$$\approx 0(1) \delta r^2 \nabla^2 V \qquad (3.72)$$

Remembering that $V = e^2/r$ where $r \sim 10^{-8} cm$, from (3.72) it follows that if $\delta r \sim l$, the Compton wavelength then

$$\frac{\Delta \alpha}{\alpha} \sim 10^{-5} \qquad (3.73)$$

where $\Delta \alpha$ is the change in the fine structure constant from the early Universe. (3.73) is an equivalent form of (3.71) (Cf.ref.[239]), and is the result originally obtained by Webb et al (Cf.refs.[237, 238]).
viii) The latest observations of distant supernovae referred to above indicate that the closure parameter $\Omega \leq 1$.
Remembering that Ω is given by [204]

$$\Omega = \frac{8\pi G}{3H^2} \rho$$

we get therefrom on using (3.2)

$$\frac{H^2}{2G} R^3 = mN$$

which immediately leads to the mysterious Weinberg formula (3.8).
ix) In General Relativity as well as in the Newtonian Theory, we have, without a cosmological constant

$$\ddot{R} = -\frac{4}{3} \pi G \rho R \qquad (3.74)$$

We remember that there is an uncertainty in time to the extent of the Compton time τ, and also if we now use the fact that G varies with time, (3.74) becomes on using (3.65),

$$\ddot{R} = -\frac{4}{3} \pi G(T - \tau) \rho R$$

$$= -\frac{4}{3} \pi G \rho R + \frac{4}{3} \pi \rho R \left(\frac{\tau}{T} \right) G \qquad (3.75)$$

Remembering that at any point of time, the age of the Universe, that is T itself is given by (3.5), we can see from (3.75) that this effect of time variation of G, which again is due to the background Zero Point Field is the same as an additional density, the vacuum density given by

$$\rho_{vac} = \frac{\rho}{\sqrt{N}} \qquad (3.76)$$

Alternatively, this term in (3.75) is also equivalent to the presence of a cosmological constant Λ. On the other hand, we know independently that the presence of a vacuum field leads to a cosmological constant given by (Cf.ref.[130] and references therein)

$$\Lambda = G\rho_{vac} \tag{3.77}$$

Equation (3.77) is pleasingly in agreement with (3.75) and (3.76) that is, the preceding considerations of fluctuational creation: In fact, due to fluctuational creation ρ_{vac} should be given by

$$\rho_{vac} = \sqrt{N}m/R^3,$$

as $\sqrt{N}$ particles are created. This gives, on using (9),

$$\rho_{vac} = \frac{m}{l^3 N} = \rho/\sqrt{N},$$

which is (3.76).

In other words quantitatively we have reconfirmed that it is the background Zero Point Field that manifests itself as the cosmological constant described in Section 3.2. This also gives as pointed out an explanation for the so called cosmological constant problem [188] viz., why is the cosmological constant so small rather than being very large.

x) In the above cosmology of fluctuations, our starting point was the creation of $\sqrt{N}$ particles within the minimum time interval, a typical elementary particle Compton time τ. We can look upon this in another way. It is well known that energy conservation can be violated, by the Uncertainty Principle, within the Compton time τ. Thus an energy $+E$ or $-E$ can appear-—the latter denoting the disappearance of energy. The nett energy dispersion for the N particles of the universe would be $\sqrt{N}E$-—this denoting the creation of $\sqrt{N}$ particles. Indeed, as we saw and will see, there as $\bar{N} \sim 10^{120}$ Planck particles that are created (Cf.equation (3.63). So $E = m_P c^2$. So the total energy this manifests is $\sqrt{\bar{N}}m_P c^2$ which gives correctly the mass energy of the universe, viz., $M = 10^{55} gm$:

$$Mc^2 = \sqrt{\bar{N}}m_P c^2$$

In this picture, as indeed we saw in the previous Chapter, our physical universe-—spacetime and matter-—is the dispersion from a larger backdrop of dark energy. A rationale for this, very much in the spirit of the condensation of particles from a background Zero Point Field as discussed at the beginning of Section 3.2, can also be obtained in terms of a phase transition from the Zero Point Field or Quantum vacuum as we will see in

the sequel. In this case, particles are like the Benard cells which form in fluids, as a result of a phase transition. While some of the particles or cells may revert to the Zero Point Field, on the whole there is a creation of $\sqrt{N}$ of these particles. If the average time for the creation of the $\sqrt{N}$ particles or cells is τ, then at any point of time where there are N such particles, the time elapsed, in our case the age of the Universe, would be given by (3.5). While this is not exactly the Big Bang scenario, there is nevertheless a rapid creation of matter from the background Quantum vacuum or Zero Point Field. Thus over 10^{40} particles would have been created within a fraction of a second.
In any case when $\tau \to 0$, we recover the Big Bang scenario with a singular creation of matter, while when $\tau \to$ Planck time we recover the Prigogine Cosmology (Cf.[130] for details). However in neither of these two limits we can deduce all the above consistent with observation Large Number relations with the cosmological constant Λ, which therefore have then to be branded as accidents.
xi) The above cosmological model is related to the fact that there are minimum space time intervals l, τ. Indeed in this case as we saw in the previous Chapter, there is an underlying non commutative geometry of spacetime given by

$$[x,y] \approx 0(l^2), [x,p_x] = \imath\hbar[1+\beta l^2], [t,E] = \imath\hbar[1+\gamma\tau^2] \tag{3.78}$$

Interestingly (3.78) implies as we saw, modification to the usual Uncertainty Principle. (This in turn has also been interpreted in terms of a variable speed of light cosmology [240–242]).
The relations (3.78), leads to the modified Uncertainty relation

$$\Delta x \sim \frac{\hbar}{\Delta p} + \alpha' \frac{\Delta p}{\hbar} \tag{3.79}$$

To see how this follows, in a simple way, we note that (3.78) implies that h goes over to $h' = h(1+\beta l^2)$. We then get (3.79) from the usual Uncertainty Principle. (3.79) appears also in Quantum Super String Theory and is related to the well known Duality relation

$$R \to \alpha'/R$$

(Cf.[243, 52]). The interpretation here is that as we go down to very small scales, we end up at the large scale as in a wormhole or a Klein's bottle. In any case (3.79) is symptomatic of the fact that we cannot go down to arbitrarily small space time intervals. That is, there is a minimum cut off. We now observe that the first term of (3.79) gives the usual Uncertainty relation. In the second term, we write $\Delta p = \Delta N mc$, where ΔN is the Uncertainty in the number of particles, in the Universe. Also $\Delta x = R$, the radius of the Universe where

$$R \sim \sqrt{N} l,$$

the famous Eddington relationship (3.9). It should be re-emphasized that the otherwise empirical Eddington formula, arose quite naturally in our cosmology.

We now get back,

$$\Delta N = \sqrt{N}$$

This is the uncertainty in the particle number, we used earlier. Substituting this in the time analogue of the second term of (3.79), we immediately get, T being the age of the Universe,

$$T = \sqrt{N} \tau$$

which is equation (3.5). So, our cosmology is self consistent with the modified relation (3.79). The fluctuational effects are really couched in the modification of the Heisenberg Principle, as given in (3.79). To put it another way, the extra term in (3.79) refers to the large scale universe and the uncertainty in the momenta, extension, energy and time spread at this scale in the sense of spacetime in the previous Chapter.

Interestingly these minimum space time considerations can be related to the Feynmann-Wheeler Instantaneous Action At a Distance formulation seen earlier, a point which we shall elaborate further in the sequel.

We finally remark that relations like (3.78) and (3.79), which can also be expressed in the form, a being the minimum length,

$$[x, p_x] = \imath\hbar[1 + \left(\frac{a}{\hbar}\right)^2 p^2]$$

(and can be considered to be truncated from a full series on the right hand side (Cf. [244])), could be deduced from the rather simple model of a lattice−−a one dimensional lattice for simplicity. In this case we will have (Cf.[130])

$$[x, p_x] = \imath\hbar cos\left(\frac{p}{\hbar}a\right),$$

where a is the lattice length, l the Compton length in our case. We will revisit a generalized version later. The energy time relation now leads to a correction to the mass energy formula, viz [245]

$$E = mc^2 cos(kl), k \equiv p/\hbar$$

This is the contribution of the extra term in the Uncertainty Principle and we will return to it in a later Chapter, in the context of observational tests.
xii) As noted the Planck scale is an absolute minimum scale in the Universe. Later we will argue that with the passage of time the Planck scale would evolve to the present day elementary particle Compton scale. To recapitulate: We have by definition

$$\hbar G/c^3 = l_P^2$$

where l_P is the Planck length $\sim 10^{-33} cms$. If we use G from (3.11) in the above we will get

$$l = N^{1/4} l_P \tag{3.80}$$

Similarly we have

$$\tau = N^{1/4} \tau_P \tag{3.81}$$

In (3.80) and (3.81) l and τ denote the typical elementary particle Compton length and time scale, and N is the number of such elementary particles in the Universe.
We could explain these equations in terms of the Benard cell like elementary particles referred to above. This time there are a total of $n = \sqrt{N}$ Planck particles and (3.80) and (3.81) are the analogues of equations (3.5) and (3.10) in the context of the formation of such particles. Indeed as we know a Planck mass, $m_P \sim 10^{-5} gms$, has a Compton life time and also a Bekenstein Radiation life time of the order of the Planck time. These spacetime scales are much too small and we encounter much too large energies from the point of view of our experimental constraints. (Moreover, the Planck scale in addition, does not show up Quantum Mechanical spin.) As noted in the previous chapter our observed scale is the Compton scale, in which Planck scale phenomena are moderated. In any case it can be seen from the above that as the number of particles N increases, the scale evolves from the Planck to the Compton scale via $n = \sqrt{N}$.
So, the scenario which emerges is, that as the Universe evolves, Planck particles form the underpinning for elementary particles, which in turn form the underpinning for the Universe by being formed continuously.
This can be confirmed by the following argument: We can rewrite (3.80) and (3.81) as

$$l = \sqrt{n} l_P, \tau = \sqrt{n} \tau_P, \tag{3.82}$$

$$l_P^2 = \frac{\hbar}{m_P} \tau_P$$

The last equation is the analogue of the diffusion process seen in the last Chapter, which is in fact the underpinning for equations like (3.5) or (3.10), except that this time we have the same Brownian process operating from the Planck scale to the Compton scale, instead of from the Compton scale to the edge of the Universe as seen above (Cf. also [157, 130]).
Interestingly, let us apply the above scenario of $\sqrt{n}$ Planck particles forming an elementary particle, to the extra term of the modified Uncertainty Principle (3.79), as we did earlier in this section. Remembering that $\alpha' = l_P^2$ in the theory, and $\Delta p = \sqrt{n} m_P c = N^{1/4} m_P c$, in this case, we get back, as $\Delta x = l$, (3.82) which is the same as (3.80) itself! Thus once again we see how the above cosmology is consistently tied up with the non commutative space time expressed by equations (3.78) or (3.79).
It may be mentioned that, as indeed can be seen from (3.80) and (3.81), in this model, the velocity of light remains constant.
xiii) We would now like to comment further upon the Compton scale and the fluctuational creation of particles alluded to above. In this case particles are being produced out of a background Quantum vacuum or Zero Point Field which is pre space time. First a Brownian process alluded to above defines the Planck length while a Brownian random process with the Planck scale as the fundamental interval leads to the Compton scale (Cf. also ref.[158]).
This process can also be modelled as a phase transition, a critical phenomenon. To see this briefly, let us start with the Landau-Ginsburg equation [246]

$$-\frac{\hbar^2}{2m}\nabla^2\psi + \beta|\psi|^2\psi = -\alpha\psi \tag{3.83}$$

Here $\hbar$ and m have the same meaning as in usual Quantum theory. It is remarkable that the above equation (3.83) is identical with a similar Schrödinger like equation based on amplitudes where moreover $|\psi|^2$ is proportional to the mass (or density) of the particle (Cf. ref.[130] for details). The equation in question is,

$$\imath\hbar\frac{\partial\psi}{\partial t} = \frac{-\hbar^2}{2m'}\frac{\partial^2\psi}{\partial x^2} + \int \psi^*(x')\psi(x)\psi(x')U(x')dx', \tag{3.84}$$

The equation (3.84) is a generalization of a two state equation deduced by Feynman a long time ago (Cf.refs.[130, 16]). We saw this in the previous Chapter. If C_1 and C_2 are the probability amplitudes for a system to be in either of two states then, we have

$$\imath\hbar\frac{dC_1}{dt} = H_{11}C_1 + H_{12}C_2 \tag{3.85}$$

$$\imath\hbar\frac{dC_2}{dt} = H_{21}C_1 + H_{22}C_2 \tag{3.86}$$

leading to two stationary states of energies $E - A$ and $E + A$, where $E \equiv H_{11} = H_{22}, A = H_{12} = H_{21}$. We can choose our zero of energy such that $E = 2A$. Indeed as has been pointed out by Feynman, when this consideration is applied to the hydrogen molecular ion, the fact that the electron has amplitudes C_1 and C_2 of being with either of the hydrogen atoms, manifests itself as an attractive force which binds the ion together, with an energy of the order of magnitude $A = H_{12}$.
The generalization of (3.85 or (3.86) is

$$\imath\hbar\frac{dC_\imath(t)}{dt} = \sum_j H_{\imath j}(t)C_j(t) \tag{3.87}$$

where the matrix $H_{\imath j}(t)$ is identified with the Hamiltonian operator. A further generalization of (3.87) leads to (3.84) if we remember that considering the continuum,

$$H(x, x') = < \psi(x)|\psi(x') >$$

In (3.84), $\psi(x)$ is the probability amplitude of a particle being at the point x and the integral is over a region of the order of the Compton wavelength. From this point of view, the similarity of (3.84) with (3.83) need not be surprising considering also that near critical points, due to universality diverse phenomena like magnetism or fluids share similar mathematical equations. Equation (3.84) was shown to lead to the Schrödinger equation with the particle acquiring a mass (Cf.also ref.[247]).
Infact in the Landau-Ginsburg case the coherence length is given by

$$\xi = \left(\frac{\gamma}{\alpha}\right)^{\frac{1}{2}} = \frac{h\nu_F}{\Delta} \tag{3.88}$$

which can be easily shown to reduce to the Compton wavelength (Cf. also ref.[248]).
Thus the emergence of Benard cell like elementary particles from the Quantum vacuum mimics the Landau-Ginsburg phase transition. In this case we have a non local growth of correlations reminiscent of the standard inflation theory.
As is known, the interesting aspects of the critical point theory (Cf.ref.[249]) are universality and scale. Broadly, this means that diverse physical phenomena follow the same route at the critical point, on the one hand, and on the other this can happen at different scales, as exemplified for example, by the course graining techniques of the Renormalization Group [250]. To

highlight this point we note that in critical point phenomena we have the reduced order parameter $\bar{Q}$ (which gives the fraction of the excess of new states) and the reduced correlation length $\bar{\xi}$ (which follows from (3.88)). Near the critical point we have relations [251] like

$$(\bar{Q}) = |t|^{\beta}, (\bar{\xi}) = |t|^{-\nu}$$

Whence

$$\bar{Q}^{\nu} = \bar{\xi}^{\beta} \tag{3.89}$$

In (3.89) typically $\nu \approx 2\beta$. As $\bar{Q} \sim \frac{1}{\sqrt{N}}$ because $\sqrt{N}$ particles are created fluctuationally, given N particles, and in view of the fractal two dimensionality of the path

$$\bar{Q} \sim \frac{1}{\sqrt{N}}, \bar{\xi} = (l/R)^2 \tag{3.90}$$

This gives back the Eddington formula,

$$R = \sqrt{N}l$$

which is nothing but (3.10).
There is another way of looking at this. The noncommutative geometry (3.78) brings out the primacy of the Quantum of Area. Indeed this has been noted from the different perspective of Black Hole Thermodynamics too [252]. We would also like to point out that a similar treatment can be easily shown to lead from the Planck scale to the Compton scale. We will return to this in a later Chapter.
In other words the creation of particles is the result of a critical point phase transition and subsequent coarse graining (Cf. also ref.[248]). So there are two equivalent models, which we have just seen. One is via a Brownian process and the second term of the modified Uncertainty Principle (3.79). This was related to large scale effects involving subconstituents like elementary particles or Planck particles. The other is through a Landau-Ginzburg process.
The above model apart from mimicking inflation also explains as we saw, the so called miraculous Large Number coincidences.
The peculiarity of these relations as noted is that they tie up large scale parameters like the radius or age of the Universe or the Hubble constant with microphysical parameters like the mass, charge and the Compton scale of an elementary particle and the gravitational constant. That is, the Universe appears to have a Machian or holistic feature. We will see that one way to understand why the large and the small are tied up is that there

is an underpinning of normal mode Planck oscillators, that is, collective phenomena all across the Universe.
For another perspective, let us go over our development. There are two types of fluctuations. One is a large scale statistical one, given by (2.1), as we saw in the previous Chapter. Here R, N and l are general. The other is the fluctuation we encounter in Quantum Mechanics−−these are at the Compton scale, l. By identifying the large scale Brownian steps l with the Compton length, we are effectively considering Quantum effects to be a result of the large scale statistical (or thermodynamic) effects. If now N is made to tend to ∞, in the thermodynamic limit, then the consequence is that $l = R/\sqrt{N} \to 0$ and so also τ. We return to the Big Bang scenario as noted and so also the Quantum Mechanical canonical commutation in (3.78) vanish and we are back with Classical Physics.
We will return to this point soon, but to re-emphasize: It has been known that there is a deep connection between a stochastic and Brownian behaviour on the one hand and critical point phenomena and the Renormalization Group on the other hand. Fractality itself is a manifestation of resolution dependent measurements, while Renormalization Group considerations arise due to coarse graining at different resolutions. A good example of the fractal behaviour is Quantum Mechanics itself which as noted earlier has been shown to have the fractal dimension 2.

3.10 Further Considerations

1. We will now provide yet another rationale for (3.4), which was our starting point. Let us start with equations encountered earlier, viz., (3.5), (3.7), (3.8), (3.9) or (3.10). For example

$$R = \sqrt{N}l$$

$$\frac{Gm^2}{e^2} = \frac{1}{\sqrt{N}} \sim 10^{-40}$$

or the Weinberg formula

$$m = \left(\frac{H\hbar^2}{Gc}\right)^{\frac{1}{3}}$$

where $N \sim 10^{80}$ is the number of elementary particles, typically pions, in the Universe. All except the last are interrelated. The very mysterious feature of (3.7) was stressed by Weinberg as we saw "...it should be noted

that the particular combination of $\hbar, H, G$, and c appearing (in the formula) is very much closer to a typical elementary particle mass than other random combinations of these quantities....
In contrast, (the formula) relates a single cosmological parameter, H, to the fundamental constants $\hbar, G, c$ and m, and is so far unexplained..."
We will now take a different route and provide an alternative theoretical rationale for these equations, and in the process light will be shed on the new cosmological model and the nature of gravitation.
Following Sivaram [253] we consider the gravitational self energy of the pion. This is given by

$$\frac{Gm^2}{l} = Gm^2/(\hbar/mc)$$

If this energy were to have a life time of the order of the age of the Universe, T, then we have by the Uncertainty relation

$$\left(\frac{Gm^3c}{\hbar}\right)(T) \approx \hbar \tag{3.91}$$

As $T = \frac{1}{H}$, this immediately gives us the Weinberg formula. It must be observed again that (3.91) gives the same time dependent gravitational constant G.
We could also derive (3.8) by using a relation given by Landsberg [254]. We use the fact that the mass of a particle is given by

$$m(b) \sim \left(\frac{\hbar^3 H}{G^2}\right)^{1/5} \left(\frac{c^5}{\hbar H^2 G}\right)^{b/15} \tag{3.92}$$

where b is an unidentified constant. Whence we have

$$m(b) \sim G^{-3/5} G^{-3b/15} = G^{-(b+1)/5}$$

The mass that would be time independent, if G were time dependent would be given by the value

$$b = -1$$

With this value of b (3.92) gives back (3.8). This provides another justification for treating m as a microphysical (constant) parameter.
Let us now proceed along a different track. We rewrite (3.91) as

$$G = \frac{\hbar^2}{m^3 c} \cdot \frac{1}{T} \tag{3.93}$$

If we use the fact that $R = cT$, then (3.93) can be written as

$$G = \frac{\hbar^2}{m^3 R} \tag{3.94}$$

Let us now use the well known relation encountered earlier viz., equation (3.3) [44]

$$R = \frac{GM}{c^2} \tag{3.95}$$

We have already seen a demonstration of (3.95) or (3.3). For another derivation, we note that the equation follows if we equate the gravitational potential energy of the pion in a three dimensional isotropic sphere of pions of radius R with the rest energy of the pion. This gives,

$$\frac{GmM}{R} = mc^2,$$

which is (3.95). If we use (3.95) in (3.94) we will get

$$G^2 = \frac{\hbar^2 c^2}{m^3 M} \tag{3.96}$$

Let $M/m = N$ be called the number of elementary particles in the Universe. In fact this is just (3.1). Then (3.96) can be written as (3.10),

$$G = \frac{\hbar c}{m^2 \sqrt{N}}$$

which can also be written as (3.8)

$$Gm^2/e^2 \sim \frac{1}{\sqrt{N}}$$

Whence we get (3.10)

$$\sqrt{N} l = R$$

We now remark that (3.93) shows an inverse dependence on time of the gravitation constant, while (3.10) shows an inverse dependence on $\sqrt{N}$. Equating the two, we get back,

$$T = \sqrt{N} \tau$$

the relation (3.5) which we have encountered several times. If we now take the time derivative of (3.11) and use (3.5), we get back

$$\dot{N} = \frac{\sqrt{N}}{\tau}$$

To put it briefly in a phase transition from the Quantum vacuum $\sqrt{N}$ particles appear within the Compton time τ. In terms of our unidirectional concept of time, we could say that particles appear and disappear, but the nett result is the appearance of $\sqrt{N}$ particles.

We now make a few remarks. Firstly it is interesting to note that $\sqrt{N} m$

will be the mass added to the Universe. Let us now apply the well known Beckenstein formula for the life time of an arbitrary mass M viz., [44],

$$t \approx G^2 M^3 / \hbar c^4$$

to the above mass. The life time as can be easily verified turns out to be exactly the age of the Universe!

A final remark. To appreciate the role of fluctuations in the otherwise mysterious Large Number relations, a feature that escaped Dirac, let us as noted following Hayakawa [126] consider the excess of electric energy due to the fluctuation $\sim \sqrt{N}$ of the elementary particles in the Universe and equate it to the inertial energy of an elementary particle. We got

$$\frac{\sqrt{N} e^2}{R} = mc^2$$

This gives us back electromagnetism-gravitation ratio if we use (3.95). If we use the Eddington formula on the other hand, we get

$$e^2 / mc^2 = l,$$

another well known relation from micro physics−−the so called classical electron radius.

2. We have noted that a background ZPF of the kind we have been considering can explain the Quantum Mechanical spin half as also the anomalous $g = 2$ factor for an otherwise purely classical electron [255, 256, 184]. The key point here is (Cf.ref.[255]) that the classical angular momentum $\vec{r} \times m\vec{v}$ does not satisfy the Quantum Mechanical commutation rule for the angular momentum $\vec{J}$. However when we introduce the background Zero Point Field, the momentum now becomes

$$\vec{J} = \vec{r} \times m\vec{v} + (e/2c)\vec{r} \times (\vec{B} \times \vec{r}) + (e/c)\vec{r} \times \vec{A}^0, \tag{3.97}$$

where $\vec{A}^0$ is the vector potential associated with the ZPF and $\vec{B}$ is an external magnetic field introduced merely for convenience, and which can be made vanishingly small.

It can be shown that $\vec{J}$ in (3.97) satisfies the Quantum Mechanical commutation relation for $\vec{J} \times \vec{J}$. At the same time we can deduce from (3.97)

$$\langle J_z \rangle = -\frac{1}{2} \hbar \omega_0 / |\omega_0| \tag{3.98}$$

Relation (3.98) gives the correct Quantum Mechanical results referred to above.

From (3.97) we can also deduce that

$$l = \langle r^2 \rangle^{\frac{1}{2}} = \left(\frac{\hbar}{mc} \right) \tag{3.99}$$

Equation (3.99) shows that the mean dimension of the region in which the ZPF fluctuation contributes is of the order of the Compton wavelength of the electron as we had noted earlier in the previous Chapter. By relativistic covariance (Cf.ref.[256]), the corresponding time scale is at the Compton scale. Thus once again we return to the Compton scale, as at the beginning of this Chapter.

3. In the light of the preceding considerations, let us now investigate the neutrino and weak interactions. We start by following Hayakawa [126] to balance the gravitational force and the Fermi energy of the "cold" background neutrinos and further identify it with the intrinsic energy of the neutrinos to get

$$\frac{GN_\nu m_\nu^2}{R} = \frac{N_\nu^{2/3}\hbar^2}{m_\nu R^2} = m_\nu c^2 \tag{3.100}$$

(All this is in the Large Number sense) m_ν is the neutrino mass. From (3.100) we can immediately deduce that

$$m_\nu = 10^{-8}m_e, N_\nu \sim 10^{90} \tag{3.101}$$

Both the relations in (3.101) are known to be correct. Indeed this mass of the neutrino was predicted by the author before the Super Kamiokande experiment threw it up in 1999 [42].

We then use the fact that due to the fluctuation in the number of neutrinos, we have an energy which is the inertial energy again:

$$\frac{\bar{g}^2\sqrt{N_\nu}}{R} \approx m_\nu c^2 \tag{3.102}$$

where $\bar{g}^2$ gives the weak interaction coupling constant.

Interestingly we have just seen a similar relation for the electrons

$$\frac{e^2\sqrt{N}}{R} = mc^2 \tag{3.103}$$

From (3.102) and (3.103) on using (3.101) we get

$$\bar{g}^2/e^2 \sim 10^{-13} \tag{3.104}$$

which ofcourse is again known to be correct.

We have thus recovered from theory the well known values of the weak coupling constant and the neutrino mass. We would next like to show that there is a complete parallel between the Large Number Relations for elementary particles with similar relations for the neutrino. We start with the simplest relation, which can be easily verified

$$N_\nu m_\nu = Nm = M = 10^{55}gm,$$

M being the mass of the Universe. We next return to the fact used above in (3.100) and consider the equality of the gravitational mass of a particle due to the remaining n particles with the inertial mass of the particle

$$\frac{Gnm^2}{r} = mc^2 \quad (3.105)$$

In (3.105), if n is replaced by N and r is replaced by the radius of the Universe, we get the mass of an elementary particle like the pion. On the other hand if in (3.105) we replace n by the number of neutrinos N_ν instead of N then we recover the mass of the neutrino. Finally if we take $n = 1$ and $r = l_P$, the Planck scale we recover the Planck mass m_P, which indeed is to be expected because as Rosen had shown and we saw earlier, the Planck mass black hole is a Universe in itself [257].
Similarly we see the complete parallel between (3.102) and (3.103). To proceed further we consider (3.10) in an alternative form viz.,

$$\hbar = \frac{Gm^2\sqrt{N}}{c} \quad (3.106)$$

For the neutrino number and neutrino mass given in (3.101), (3.106) gives

$$\hbar' = \frac{Gm_\nu^2\sqrt{N_\nu}}{c} = 10^{-12}\hbar \quad (3.107)$$

(3.107) shows that the magnetic moment of the neutrino is given by

$$\mu_\nu \sim 10^{-11}\,\text{Bohr} \quad \text{magnetons} \quad (3.108)$$

Indeed (3.108) is consistent with observation [258]. That is for the neutrino we have effectively $\hbar'$ given by (3.107), instead of $\hbar$. It is then simple to verify that the analogue of the Eddington formula (9) applies for the neutrinos viz.,

$$R = \sqrt{N_\nu}l_\nu,$$

where $l_\nu = \frac{\hbar'}{m_\nu c}$, the neutrino analogue of the Compton length. Interestingly, this neutrino "Compton" wavelength in $\sim 10^{-15}cm$, corresponding roughly to the mass of a W boson, the carrier of weak interaction.
It has been shown on the basis of black hole radiation life times that we have

$$\frac{Gm^2}{l} = \frac{\hbar}{T}, T = 10^{17}sec \quad (3.109)$$

where T is the life time of the Universe (Cf. also [253]). Indeed as we saw (3.109) is just a variant of the Weinberg formula, and can now be interpreted as the fact that the gravitational self energy of the elementary particle, viz., $\frac{Gm^2}{l}$ has a life time of the order of the age of the Universe, due to the Uncertainty Principle. It can immediately be verified that for the neutrino we have the equation

$$\frac{Gm_\nu^2}{l_\nu} = \frac{\hbar'}{T} \tag{3.110}$$

In the author's model, it has been shown that [130] the pion can be considered to be an electron positron bound state so that we have

$$l = \frac{e^2}{m_e c^2} \tag{3.111}$$

where l is the pion Compton wavelength. Equation (3.111) is on a different footing from an electron-positron pair annihilating itself. This is because in our picture time (and space) is no longer continuous. Similarly one could consider the pion to also be the bound state of a quark anti-quark in QCD so that we have

$$\frac{g^2}{m_q c^2} = l \tag{3.112}$$

where m_q is the quark mass and g^2 is the strong interaction coupling constant. There is an immediate analogue of (3.111) and (3.112) for the neutrino viz.,

$$l_\nu = \frac{\bar{g}^2}{m_\nu c^2} \tag{3.113}$$

Finally it may be pointed out that there is an immediate analogue of the Weinberg formula (7) as well, viz.,

$$m_\nu = \left(\frac{H\hbar'^2}{Gc}\right)^{1/3} \tag{3.114}$$

It must be mentioned that these analogues like (3.102), (3.107), (3.110), (3.113) and (3.114) between the neutrino and an elementary particle are not mere numerical coincidences. This is because the various relations for the elementary particles are the result of a theoretical structure, and are not mere accidents. What the foregoing means is that the neutrino has a similar theoretical structure.

4. Remaining with the neutrino background, we can show that this too provides us with a physical mechanism for the cosmological constant. Let us first treat the cold cosmic neutrino background as a degenerate Fermi assembly. We have [125]

$$p_F^2 = \hbar^3(N/V) \tag{3.115}$$

Feeding in the known neutrino parameter, viz., [259] $N_\nu \sim 10^{90}$ we get from the above, the correct neutrino mass $\sim 10^{-3}eV$ and the background temperature $T \sim 1°K$. More recently there has been hope that neutrinos can also exhibit the ripples of the early Big Bang and in fact, Trotta and Melchiorri claim to have done so [260].

It may be mentioned that there is growing evidence for the cosmic background neutrinos [261]. The GZK photo pion process to which we will return in a later Chapter, seems to be the contributing factor.

Let us now consider this neutrino background to deduce the correct cosmological constant. We note that the cosmological constant is given by

$$\lambda = < 0|H|0 > \equiv \text{cosmological} \quad \text{constant} \tag{3.116}$$

The cosmological constant λ is now given by its familiar expression [26]

$$\lambda = \int_0^\Lambda \frac{4\pi p^2}{(2\pi)^3} dp \frac{1}{2} \sqrt{p^2 + m^2} \tag{3.117}$$

In (3.117) Λ is the cut off which takes care of the divergent integral. If we now use the value of the neutrino mass $\sim 10^{-3}eV$ in (3.117) then we get the value of the cosmological constant as

$$\lambda \sim 10^{-50} GeV^4 \tag{3.118}$$

which is consistent with the latest observations pertaining to the accelerating universe with a small cosmological constant.

On the other hand, in the usual theory, Λ has been taken to correspond to the Planck scale and the Planck mass $\sim 10^{19}GeV$. This has lead to the value of the cosmological constant which is 10^{120} times its actual value. We have already encountered the famous cosmological constant problem. On the contrary, in (3.117) we could consider the photon mass $\sim 10^{-65}gms$ which we have alluded to. This gives a contribution many orders of magnitude below the correct value given in (3.118)−−as such the contribution of the photon background is negligible.

We can now see that by considering the cosmic neutrino background rather than the Planck cut off, we get the right order of the cosmological constant. Further references to the cosmological constant may be found in [209, 263–265] (and references therein).

5. We finally observe that it is not surprising that Quantum Theory should be the effect of fluctuations in the universe as a whole. In fact as pointed out [126] the fluctuation in the mass of a typical elementary particle, for

example the pion, due to the fluctuation $\sim \sqrt{N}$ of the particle number $N \sim 10^{80}$ is given by (Cf.ref.[198])

$$\Delta mc^2 \approx \frac{G\sqrt{N}m^2}{R}$$

As this energy is proportional inversely to the time period, which in this case of the age of the universe, T, we have

$$\beta = T \cdot \Delta mc^2 = G\sqrt{N}m^2T/R \tag{3.119}$$

where β in (3.119) is the constant of proportionality. We can easily verify that $\beta \sim \hbar$ if $T = cR$. In fact Equation (3.119) itself is an expression of the uncertainty relation

$$\Delta E \Delta t \approx \hbar$$

So Equation (3.119) suggests the origin of Quantum Theory in cosmic fluctuations.

Inaba [266] deduces for a nearly flat Robertson-Walker universe from a minimum average curvature principle, the Hamilton-Jacobi equation for a single particle,

$$\partial_t S + \frac{1}{2m} g^{\imath\jmath}(\nabla, S) - xR = 0 \tag{3.120}$$

where the curvature R is given by

$$R = R^{(b)} - R' : R^{(b)} = 6\left(\frac{\dot{a}}{a} + \frac{\dot{a}^2}{a^2}\right)$$

R' being the fluctuation effect and $R^{(b)}$ being the curvature in the standard Robertson-Walker geometry.

Equation (3.120) leads by the standard Madelung-Bohm or Nelson theory to the Schrodinger equation

$$\imath\hbar\partial\psi = \frac{\hbar^2}{2m}\Delta\psi + V\psi - \frac{\hbar^2}{4m}R^{(b)} \tag{3.121}$$

We can then argue that (3.121) is indeed the Quantum Mechanical equation in the classical Robertson-Walker geometry. It is the perturbation R' in the Robertson-Walker geometry that leads to (3.121). Also we are able to see the origin of the mysterious Quantum potential term V in the above equation—–it comes for the fluctuation R' in the curvature.

We can justify the above conclusion as follows: We have already observed in the previous Chapter and earlier in this Chapter that in the random motion of N particles the fluctuation in the length is given by (3.10).

Further as we saw, the diffusion equation describing the motion of a particle with position given by $x(t)$ subject to random corrections is given by the well-known equation

$$|\Delta| = \sqrt{\langle \Delta x^2 \rangle} \approx \nu \sqrt{\Delta t}$$

where the diffusion constant ν is related to the mean free path l and the mean velocity v

$$\nu \approx lv \qquad (3.122)$$

Identifying l of Equation (3.122) with that in (3.10) we arrive at the Hamilton-Jacobi equation (3.120) and thence the Schrodinger equation (3.121) [130, 127, 176, 267].
Thus using the equations of Brownian motion in the context of all the particles in the universe, we arrive at the equations (3.120) and (3.121) based on a minimum curvature principle and Santamato's geometric Quantum Mechanics which too we will briefly encounter later.
In fact one can look upon the above results in terms of the fluctuation of the metric itself. In Santamato's original formulation [268]-[271] the geometry is Weyl's gauge invariant geometry, where there is no invariant length and in fact we have

$$\delta l^2 \sim l^2 \delta g_{\imath k} \qquad (3.123)$$

It must be stressed that (3.123) is valid for arbitrary vectors A^μ in which case l would be their length.
Using the usual geometrodynamic formula for the fluctuation of the metric [45] we have

$$l^2 \delta g_{\imath k} \approx \frac{l_P}{l}, \qquad (3.124)$$

where l_P is the Planck length.
Whence we get

$$\delta g_{\imath k} \sim l \qquad (3.125)$$

if l is of the order $10^{-11} cm$ or the electron Compton wavelength.
Similarly using (3.10) in (3.123), we recover (3.125), as in the Weyl geometry. Finally it may be mentioned that even from the usual geometrodynamic point of view, the fluctuations in the curvature over a region of length a are given by [45],

$$\Delta r \sim l_P / a^3,$$

where r is the curvature. In the macro world this is small, but if a is the Compton wavelength, then $\Delta r \geq 1$. Thus at this scale, Quantum effects come in.

This establishes the equivalence of the two approaches and reconfirms the cosmic, Machian feature, from a more general viewpoint. This apart it provides a rationale for the Quantum Potential term encountered in the Brownian (and Bohmian) characterization of the last Chapter. It must be reiterated that this is the only "non-classical" term in an otherwise classical development of the theory.

Chapter 4

The Thermodynamic Universe

4.1 Introduction

The view we are trying to put forward is that the Universe is in some sense immersed in a bath of "Dark Energy" and is "Thermodynamic" in the sense that for example the temperature of some material has a certain value. This value emerges from the combined statistical motion of several individual molecules whereas in a reductionist view, in this particular example, we would be determining the motion of one of these typically 10^{23} molecules. To get a flavour, let us first re-derive the recently discovered [21] residual cosmic energy directly from the background Dark Energy. We may reiterate that the "mysterious" background Dark Energy is the same as the quantum Zero Point Fluctuations in the background vacuum electromagnetic field as seen in the last Chapter. Let us recall that the background Zero Point Field is a collection of ground state oscillators [45]. The probability amplitude is

$$\psi(x) = \left(\frac{m\omega}{\pi\hbar}\right)^{1/4} e^{-(m\omega/2\hbar)x^2}$$

for displacement by the distance x from its position of classical equilibrium. So the oscillator fluctuates over an interval

$$\Delta x \sim (\hbar/m\omega)^{1/2}$$

The background electromagnetic field is an infinite collection of independent oscillators, with amplitudes X_1, X_2 etc. The probability for the various oscillators to have amplitudes X_1, X_2 and so on is the product of individual oscillator amplitudes:

$$\psi(X_1, X_2, \cdots) = exp[-(X_1^2 + X_2^2 + \cdots)]$$

wherein there would be a suitable normalization factor. This expression gives the probability amplitude ψ for a configuration $B(x,y,z)$ of the magnetic field that is described by the Fourier coefficients $X_1, X_2, \cdots$ or directly in terms of the magnetic field configuration itself by, as we saw,

$$\psi(B(x,y,z)) = Pexp\left(-\int\int\frac{\mathbf{B(x_1)}\cdot\mathbf{B(x_2)}}{16\pi^3\hbar c r_{12}^2}d^3x_1d^3x_2\right).$$

P being a normalization factor. At this stage, we are thinking in terms of energy without differenciation, that is, without considering Electromagnetism or Gravitation etc as separate. Let us consider a configuration where the magnetic field is everywhere zero except in a region of dimension l, where it is of the order of $\sim \Delta B$. The probability amplitude for this configuration would be proportional to

$$\exp[-((\Delta B)^2 l^4/\hbar c)]$$

So the energy of fluctuation in a region of length l is given by finally, the density [45, 16]

$$B^2 \sim \frac{\hbar c}{l^4}$$

So the energy content in a region of volume l^3 is given by

$$\beta^2 \sim \hbar c/l \tag{4.1}$$

This energy is minimum when l is maximum. Let us take l to be the radius of the Universe $\sim 10^{28}cms$. The minimum energy residue of the background Dark Energy or Zero Point Field (ZPF) now comes out to be $10^{-33}eV$, exactly the observed value. This observed residual energy is a cosmic footprint of the ubiquitous Dark Energy in the Universe, a puzzling footprint that, as we have noted, has recently been observed [21].
If on the other hand we take for l in (4.1) the smallest possible length, which has been taken to the Planck length l_P, as we will see in the sequel, then we get the Planck mass $m_P \sim 10^{-5}gm$.
The minimum mass $\sim 10^{-33}eV$ or $10^{-65}gms$, will be seen to be the mass of the photon. Interestingly, this also is the minimum thermodynamic mass in the Universe, as shown by Landsberg from a totally different point of view, that of thermodynamics as we saw in the previous Chapter [254]. So (4.1) gives two extreme masses, the Planck mass and the photon mass. We will see how it is possible to recover the intermediate elementary particle mass from these considerations later in this Chapter.
As an alternative derivation, it is interesting to derive a model based on the theory of phonons which are quanta of sound waves in a macroscopic body [125]. Phonons are a mathematical analogue of the quanta of the electromagnetic field, which are the photons, that emerge when this field is

expressed as a sum of Harmonic oscillators. This situation is carried over to the theory of solids which are made up of atoms that are arranged in a crystal lattice and can be approximated by a sum of Harmonic oscillators representing the normal modes of lattice oscillations. In this theory, as is well known the phonons have a maximum frequency ω_m which is given by

$$\omega_m = c\left(\frac{6\pi^2}{v}\right)^{1/3} \tag{4.2}$$

In (4.2) c represents the velocity of sound in the specific case of photons, while $v = V/N$, where V denotes the volume and N the number of atoms. In this model we write

$$l \equiv \left(\frac{4}{3}\pi v\right)^{1/3}$$

l being the inter particle distance. Thus (4.2) now becomes

$$\omega_m = c/l \tag{4.3}$$

Let us now liberate the above analysis from the immediate scenario of atoms at lattice points and quantized sound waves due to the Harmonic oscillations and look upon it as a general set of Harmonic oscillators as above. Then we can see that (4.3) and (4.1) are identical as

$$\omega = \frac{mc^2}{\hbar}$$

So we again recover with suitable limits the extremes of the Planck mass and the photon mass. (Other intermediate elementary particle masses follow if we take l as a typical Compton wavelength.)

We now examine separately, the Planck scale and the photon mass. As we saw earlier, there were basically two concepts of space which we had inherited from the early days of modern science. The predominant view has been the legacy from the Newtonian world view as we saw. Here we consider spacetime to form a differentiable manifold. We also saw that Liebniz had a different view of space, not as a container, but rather as made up of the contents itself. This lead to a view where spacetime has the smallest unit, and therefore non differentiable.

Max Planck had noticed that, what we call the Planck scale today,

$$l_P = \left(\frac{\hbar G}{c^3}\right)^{\frac{1}{2}} \sim 10^{-33} cm \tag{4.4}$$

is made up of the fundamental constants of nature and so, he suspected it played the role of a fundamental length. Indeed, modern Quantum Gravity

approaches have invoked (4.4) in their quest for a reconciliation of gravitation with other fundamental interactions. In the process, the time honoured prescription of a differentiable spacetime has to be abandoned. Later, we will try to give a rationale for (4.4) being the smallest scale.
There is also another scale made up of fundamental constants of nature, viz., the well known Compton scale (or classical electron radius which we encountered in Chapter 1),

$$l = e^2/m_e c^2 \sim 10^{-12} cm \tag{4.5}$$

where e is the electron charge and m_e the electron mass. We had seen how the Compton scale emerges from the ZPF, in the previous Chapter. This had appeared in the Classical theory of the electron unlike the Planck scale, which was a product of Quantum Theory. Indeed if (4.5) is substituted for l in (4.1), we get the elementary particle mass scale.
The scale (4.5) has also played an important role in modern physics, though it is not considered as fundamental as the Planck scale. Nevertheless, the Compton scale (4.5) is close to reality in the sense of experiment, unlike (4.4), which is well beyond foreseeable direct experimental contact. Moreover another interesting feature of the Compton scale is that, as we saw in the last Chapter, it brings out the Quantum Mechanical spin, unlike the Planck scale.
A very important question this throws up is that of a physical rationale for a route from (4.4) to (4.5). Is there such a mechanism? We have already seen one such route in the last Chapter: via phase transitions, $n \sim 10^{40}$ Planck particles "condense" into an elementary particle. Let us investigate further.

4.2 The Planck and Compton Scales

We have seen that String Theory, Loop Quantum Gravity and a few other approaches start from the Planck scale. This is also the starting point in our alternative theory of Planck oscillators in the background dark energy. We first give a rationale for the fact that the Planck scale would be a minimum scale in the Universe [272]. Our starting point [158, 16] is the model for the underpinning at the Planck scale for the Universe. This is a collection of N Planck scale oscillators where we will specify N shortly.
Earlier, we had argued in the last Chapter that a typical elementary particle like a pion could be considered to be the result of $n \sim 10^{40}$ evanescent Planck scale oscillators. We will now consider the problem from a different

point of view, which not only reconfirms the above result, but also enables an elegant extension to the case of the entire Universe itself. Let us consider an array of N particles, spaced a distance Δx apart, which behave like oscillators that are connected by springs. We then have [157, 246, 273, 16]

$$r = \sqrt{N\Delta x^2} \tag{4.6}$$

$$ka^2 \equiv k\Delta x^2 = \frac{1}{2}k_B T \tag{4.7}$$

where k_B is the Boltzmann constant, T the temperature, r the extent and k is the spring constant given by

$$\omega_0^2 = \frac{k}{m} \tag{4.8}$$

$$\omega = \left(\frac{k}{m}a^2\right)^{\frac{1}{2}} \frac{1}{r} = \omega_0 \frac{a}{r} \tag{4.9}$$

We now identify the particles with Planck masses and set $\Delta x \equiv a = l_P$, the Planck length. It may be immediately observed that use of (4.8) and (4.7) gives $k_B T \sim m_P c^2$, which of course agrees with the temperature of a Black Hole of Planck mass. Indeed, Rosen [257] had shown that a Planck mass particle at the Planck scale can be considered to be a Universe in itself with a Schwarzchild radius equalling the Planck length. We also use the fact alluded to that a typical elementary particle like the pion can be considered to be the result of $n \sim 10^{40}$ Planck masses.
Using this in (4.6), we get $r \sim l$, the pion Compton wavelength as required. Whence the pion mass is given by

$$m = m_P/\sqrt{n}$$

Further, in this latter case, using (4.6) and the fact that $N = n \sim 10^{40}$, and (4.7),i.e. $k_B T = kl^2/N$ and (4.8) and (4.9), we get for a pion, remembering that $m_P^2/n = m^2$,

$$k_B T = \frac{m^3 c^4 l^2}{\hbar^2} = mc^2,$$

which of course is the well known formula for the Hagedorn temperature for elementary particles like pions [140]. In other words, this confirms the earlier conclusions that we can treat an elementary particle as a series of some 10^{40} Planck mass oscillators.
However it must be observed from (4.9) and (4.8), that while the Planck mass gives the highest energy state, an elementary particle like the pion

is in the lowest energy state. This explains why we encounter elementary particles, rather than Planck mass particles in nature. Infact as already noted [18], a Planck mass particle decays via the Beckenstein radiation within a Planck time $\sim 10^{-42} secs$. On the other hand, the lifetime of an elementary particle would be very much higher.
In any case the efficacy of our above oscillator model can be seen by the fact that we recover correctly the masses and Compton scales in the order of magnitude sense and also get the correct Bekenstein and Hagedorn formulas as seen above, and further we even get the correct estimate of the mass and size of the Universe itself, as will be seen below.
Using the fact that the Universe consists of $N \sim 10^{80}$ elementary particles like the pions, the question is, can we think of the Universe as a collection of nN or 10^{120} Planck mass oscillators? This is what we will now show. Infact if we use equation (4.6) with

$$\bar{N} \sim 10^{120},$$

we can see that the extent is given by $r \sim 10^{28} cms$ which is of the order of the diameter of the Universe itself. We shall shortly justify the value for $\bar{N}$. Next using (4.9) we get

$$\hbar\omega_0^{(min)} \langle \frac{l_P}{10^{28}} \rangle^{-1} \approx m_P c^2 \times 10^{60} \approx Mc^2 \tag{4.10}$$

which gives the correct mass M, of the Universe which in contrast to the earlier pion case, is the highest energy state while the Planck oscillators individually are this time the lowest in this description. In other words the Universe itself can be considered to be described in terms of normal modes of Planck scale oscillators.
More generally, if an arbitrary mass M, as in (4.10), is given in terms of $\bar{N}$ Planck oscillators, in the above model, then we have from (4.10) and (4.6):

$$M = \sqrt{\bar{N}} m_P \,\text{and}\, R = \sqrt{\bar{N}} l_P,$$

where R is the radius of the object. Using the fact that l_P is the Schwarzchild radius of the mass m_P, this gives immediately,

$$R = 2GM/c^2$$

a relation we have deduced alternatively in the previous Chapter. In other words, such an object, the Universe included as a special case, shows up as a Black Hole.
We will return to these considerations later: this and the preceding considerations merely set the stage (Cf.refs.[158, 157, 16, 274, 273] for details).

In fact, we do not need to specify N. We have in this case rewriting the relations (4.6) to (4.9),

$$R = \sqrt{N}l, Kl^2 = kT,$$

$$\omega_{max}^2 = \frac{K}{m} = \frac{kT}{ml^2} \tag{4.11}$$

In (4.11), R is of the order of the diameter of the Universe, K is the analogue of spring constant, T is the effective temperature while l is the analogue of the Planck length, m the analogue of the Planck mass and ω_{max} is the frequency−−the reason for the subscript max will be seen below. We do not yet give l and m their usual values as given in (4.4) for example, but rather try to deduce these values.

We now use the well known result alluded to that the individual minimal oscillators are black holes or mini Universes as shown by Rosen [257]. So using the Beckenstein temperature formula for these primordial black holes [44], that is

$$kT = \frac{\hbar c^3}{8\pi Gm}$$

in (4.11) we get,

$$Gm^2 \sim \hbar c \tag{4.12}$$

which is another form of (4.4). We can easily verify that (4.12) leads to the value $m \sim 10^{-5}gms$. In deducing (4.12) we have used the typical expressions for the frequency as the inverse of the time−−the Compton time in this case and similarly the expression for the Compton length. However it must be reiterated that no specific values for l or m were considered in the deduction of (4.12).

We now make two interesting comments. Cercignani and co-workers have shown [275, 276] that when the gravitational energy becomes of the order of the electromagnetic energy in the case of the Zero Point oscillators, that is

$$\frac{G\hbar^2\omega^3}{c^5} \sim \hbar\omega \tag{4.13}$$

then this defines a threshold frequency ω_{max} above which the oscillations become chaotic. In other words, for meaningful physics we require that

$$\omega \leq \omega_{max}.$$

Secondly as we saw from the parallel but unrelated theory of phonons [125, 108], which are also bosonic oscillators, we deduce a maximal frequency given by

$$\omega_{max}^2 = \frac{c^2}{l^2} \tag{4.14}$$

In (4.14) c is, as we saw in the particular case of phonons, the velocity of propagation, that is the velocity of sound, whereas in our case this velocity is that of light. Frequencies greater than ω_{max} in (4.14) are again meaningless. We can easily verify that using (4.13) in (4.14) gives back (4.12).
Finally we can see from (4.11) that, given the value of l_P and using the value of the radius of the Universe, viz., $R \sim 10^{27}$, we can deduce that,

$$N \sim 10^{120} \tag{4.15}$$

In a sense the relation (4.12) can be interpreted in a slightly different vein as representing the scale at which all energy– gravitational and electromagnetic becomes one. In any case (4.15) justifies our earlier choice of $\bar{N}$. If $N \sim 10^{80}$ elementary particles in the Universe is taken as an input, then we can deduce $n \sim 10^{40}$ is the number of Planck oscillations in a typical elementary particle.
It should also be borne in mind that, a Planck scale particle is a Schwarzchild Black Hole. From this point of view, we cannot penetrate the Planck Scale−−it constitutes a physical limit.
The Compton scale (4.5) comes as a Quantum Mechanical effect, within which we have zitterbewegung effects and a breakdown of causal Physics as emphasized [15]. Indeed as we noted Dirac had noticed this aspect in connection with two difficulties with his electron equation. Firstly the speed of the electron turns out to be the velocity of light. Secondly the position coordinates become complex or non Hermitian. His explanation as we saw was that in Quantum Theory we cannot go down to arbitrarily small spacetime intervals, for the Heisenberg Uncertainty Principle would then imply arbitrarily large momenta and energies. So Quantum Mechanical measurements are actually an average over intervals of the order of the Compton scale.
Weinberg also noticed the non physical aspect of the Compton scale [17]. Starting with the usual light cone of Special Relativity and the inversion of the time order of events, he goes on to add, and we quote again at a little length and comment upon it,
"Although the relativity of temporal order raises no problems for classical physics, it plays a profound role in quantum theories. The uncertainty

principle tells us that when we specify that a particle is at position x_1 at time t_1, we cannot also define its velocity precisely. In consequence there is a certain chance of a particle getting from x_1 to x_2 even if $x_1 - x_2$ is spacelike, that is, $|x_1 - x_2| > |x_1^0 - x_2^0|$. To be more precise, the probability of a particle reaching x_2 if it starts at x_1 is nonnegligible as long as

$$(x_1 - x_2)^2 - (x_1^0 - x_2^0)^2 \leq \frac{\hbar^2}{m^2}$$

where $\hbar$ is Planck's constant (divided by 2π) and m is the particle mass. (Such space-time intervals are very small even for elementary particle masses; for instance, if m is the mass of a proton then $\hbar/m = 2 \times 10^{-14} cm$ or in time units $6 \times 10^{-25} sec$. Recall that in our units $1sec = 3 \times 10^{10} cm$.) We are thus faced again with our paradox; if one observer sees a particle emitted at x_1, and absorbed at x_2, and if $(x_1 - x_2)^2 - (x_1^0 - x_2^0)^2$ is positive (but less than or $= \hbar^2/m^2$), then a second observer may see the particle absorbed at x_2 at a time t_2 before the time t_1 it is emitted at x_1.

"There is only one known way out of this paradox. The second observer must see a particle emitted at x_2 and absorbed at x_1. But in general the particle seen by the second observer will then necessarily be different from that seen by the first. For instance, if the first observer sees a proton turn into a neutron and a positive pi-meson at x_1 and then sees the pi-meson and some other neutron turn into a proton at x_2, then the second observer must see the neutron at x_2 turn into a proton and a particle of negative charge, which is then absorbed by a proton at x_1 that turns into a neutron. Since mass is a Lorentz invariant, the mass of the negative particle seen by the second observer will be equal to that of the positive pi-meson seen by the first observer. There is such a particle, called a negative pi-meson, and it does indeed have the same mass as the positive pi-meson. This reasoning leads us to the conclusion that for every type of charged particle there is an oppositely charged particle of equal mass, called its antiparticle. Note that this conclusion does not obtain in non-relativistic quantum mechanics or in relativistic classical mechanics; it is only in relativistic quantum mechanics that antiparticles are a necessity. And it is the existence of antiparticles that leads to the characteristic feature of relativistic quantum dynamics, that given enough energy we can create arbitrary numbers of particles and their antiparticles."

In Weinberg's analysis, one observer sees only protons at x_1 and x_2, whereas the other observer sees only neutrons at x_1 and x_2 while in between, the first observer sees a positively charged pion and the second observer a negatively charged pion. In other words, the event seen by the first observer is different from that seen by the second observer in a profound way. There is no longer the "same" event, for example a proton decaying into neutron, but seen by different observers. We will come back to this point later but remark that Weinberg's explanation is in the spirit of the Feynman-Stuckleberg diagrams.

4.3 The Transition

We now address the question we raised earlier, that of the mechanism by which there is a transition from the Planck scale to the Compton scale. We have already seen this in the previous Chapter, but look at it now, from a different perspective. For this we will need a relation we encountered in the last Chapter [175, 176]

$$G = \Theta/t \tag{4.16}$$

As we saw, the relation (4.16) shows the gravitational constant as varying with time. This dependence also features in the Dirac Cosmology and a few other cosmologies as noted[43].

We now observe the following: It is known that for a Planck mass $m_P \sim 10^{-5}gm$, all the energy is gravitational and in fact we have, as in (4.12),

$$Gm_P^2 \sim e^2$$

For such a mass the Schwarzschild radius is the Planck length or Compton length for a Planck mass

$$\frac{Gm_P}{c^2} = l_P \sim \hbar/m_P c \sim 10^{-33}cm \tag{4.17}$$

To push these considerations further, we have from the theory of black hole thermodynamics [44] for any arbitrary mass m, the well known Beckenstein temperature encountered earlier and given by

$$T = \frac{\hbar c^3}{8\pi kmG} \tag{4.18}$$

We can deduce this relation (4.18), even from our Planck oscillator theory. For this we use the following relations for a Schwarzchild black hole [44]:

$$dM = TdS,\ S = \frac{kc}{4\hbar G}A,$$

where T is the Beckenstein temperature, S the entropy and A is the area of the black hole. In our case, the mass $M = \sqrt{N}m_P$ and $A = Nl_P^2$, where N is arbitrary for an arbitrary black hole. Whence,

$$T = \frac{dM}{dS} = \frac{4\hbar G}{kl_P^2 c}\frac{dM}{dN}$$

If we use the fact that l_P is the Schwarzchild radius for the Planck mass m_P and use the expression for M, the above reduces to (4.18).
Equation (4.18) gives the thermodynamic temperature of a Planck mass black hole. Further, in this theory as is known [44],

$$\frac{dm}{dt} = -\frac{\beta}{m^2}, \tag{4.19}$$

where β is given by

$$\beta = \frac{\hbar c^4}{(30.8)^3 \pi G^2}$$

This leads back to the usual black hole life time given by, for any mass m,

$$t = \frac{1}{3\beta} m^3 = 8.4 \times 10^{-24} m^3 secs \tag{4.20}$$

Let us now factor in the time variation (4.16) of G into (4.19). Equation (4.19) now becomes [18]

$$m^2 dm = -B\,\mu^{-2} t^2 dt, B \equiv \frac{\hbar c^4}{\lambda^3 \pi}, \mu \equiv \frac{l c^2 \tau}{m}, \lambda^3 = (30.8)^3 \pi$$

Whence on integration we get

$$m = \frac{\hbar}{\lambda \pi^{1/3}} \left\{ \frac{1}{l^6} \right\}^{1/3} t = \frac{\hbar}{\lambda \pi^{1/3}} \frac{1}{l^2} t \tag{4.21}$$

If we use the pion Compton time for t, in (4.21), we get for m, the pion mass. In other words, due to (4.16), the evanescent Planck mass decays into a stable elementary particle.
We can also come to this conclusion from an alternative viewpoint. We can compare (4.17) with (4.5) which defines l as what may be called the "electromagnetic Schwarzschild" radius viz., the Compton wavelength, when e^2 is seen as an analogue of Gm^2.
If now we carry out the substitution $Gm^2 \to e^2$ in the above we have instead of (4.18), the relation

$$kT \sim mc^2 \tag{4.22}$$

Equation (4.22) is the well known relation expressing the Hagedorn temperature of elementary particles [140]. Similarly instead of (4.19) we will get

$$\frac{dm}{dt} = -\frac{\hbar c^4}{\Theta^3 e^4} m^2, \Theta^3 = (30.8)^3 \pi$$

Whence we get for the life time

$$\frac{\hbar c^4}{\Theta^3 e^4} t = \frac{1}{m} \tag{4.23}$$

From (4.23) we get, for the pion, a life time

$$t \sim 10^{-23} secs,$$

which is the pion Compton time. So the Compton time shows up as an "electromagnetic Beckenstein radiation life time."

Thus for elementary particles, working within the context of gravitational theory, but with a scaled up coupling constant, we get the meaningful relations (4.22) and (4.23) giving the Compton time and Compton length as also the Hagedorn temperature as the analogues of the Schwarzschild radius, radiation life time and black hole temperature obtained with the usual Gravitational coupling constant.

Before proceeding, we observe that we have deduced that a string of N Planck oscillators, N arbitrary, form a Schwarzchild black hole of mass $\sqrt{N} m_P = M$. Using the analogue of (4.11) and considerations before (4.19), we can deduce that

$$\frac{dM}{dt} = m_P / t_P,$$

$$M = (m_P / t_P) \cdot t,$$

Whence t being the "Beckenstein decay time" this is like the equation (4.21). For the Planck mass, $M = m_P$, the decay time $t = t_P$. For the Universe, the above gives the life time t as $\sim 10^{17} sec$, the age of the Universe! Interestingly, for such black holes to have realistic life times $\geq 1 sec$, we deduce that the mass $\geq 10^{38} gm$.

A long standing puzzle has been the so called Gauge Hierarchy Problem. This deals with the fact that the Planck mass is some 10^{20} times the mass of an elementary particle, for example Gauge bosons or protons or electrons (in the large number sense). Why is there such a huge gap? In a sense, this is the same as the question, why the Planck scale is 10^{20} times smaller than the elementary particle scale [272].

We now recall that [175, 176, 277, 16]

$$G = \frac{\hbar c}{m^2 \sqrt{N}} \tag{4.24}$$

In (4.24) $N \sim 10^{80}$ is as before the number of elementary particles in the Universe.

What is interesting about (4.24) is that it shows gravitation as a distributional effect over all the N particles in the Universe [277, 16]. We will return to this important aspect in a later Chapter.
Let us rewrite (4.17) in the form

$$G \approx \frac{\hbar c}{m_P^2} \tag{4.25}$$

remembering that the Planck length is also the Compton length of the Planck mass. (Interestingly an equation like (4.17) or (4.25) also follows from Sakharov's treatment of gravitation [278] as we will see later.) A division of (4.24) and (4.25) yields

$$m_P^2 = \sqrt{N} m^2 \tag{4.26}$$

Equation (4.26) which we saw earlier as

$$m_P = \sqrt{n} m$$

where $n \sim 10^{40}$, immediately gives the ratio $\sim 10^{20}$ between the Planck mass and the mass of an elementary particle.
Comparing (4.24) and (4.25) we can see that the latter is the analogue of the former in the case $N \sim 1$. So while the Planck mass in the spirit of Rosen's isolated Universe and the Schwarzchild black hole uses the (gravitational) interaction in isolation, as seen from (4.24), elementary particles are involved in the gravitational interaction with all the remaining particles in the Universe.
Finally rememebring that $Gm_P^2 \sim e^2$, as can also be seen from (4.25), we get from (4.24)

$$\frac{e^2}{Gm^2} \sim \frac{1}{\sqrt{N}} \tag{4.27}$$

Equation (4.27) is the otherwise empirically well known Electromagnetism-Gravitation coupling constant ratio encountered earlier.
It may be remarked that one could attempt an explanation of (4.26) from the point of view of Super Symmetry or Brane theory, but these latter have as yet no experimental validation [279]. On the other hand, as seen in the previous Chapter the crucial equation (4.24) was actually part of a cosmology which predicted a dark energy driven accelerating Universe with a small cosmological constant (besides a deduction of hitherto empirical Large Number relations).

4.4 Photon Mass

We next come to the mass of the photon. As is well known the concept of the photon grew out of the work of Planck and Einstein, though its earliest origin was in Newton's Corpuscular Theory. Thereafter the photon got integrated into twentieth century physics, be it Classical or Quantum. Though it is considered to be a massless particle of spin 1 and 2 helicity states (as proved later for any massless particle with spin by Wigner), it is interesting to note that there had been different dissenting views.
De Broglie himself [280] believed that the photon has a mass, a view shared by a few others as well. (Interestingly in 1940 and 1942, De Broglie published two volumes on the Theory of Light, La mecanique ondulatoire du photon Une nouvelle theorie de la lumiere, the first volume, La lumiere dans le vide (Paris, Hermann, 1940); the second volume, Les interactions entre les photons et la matiere (Paris, Hermann, 1942) [281, 282]. An apparent objection to this view has been that a photon mass would be incompatible with Special Relativity. However it is interesting to note that nowhere in twentieth century physics has it been proved that the photon has no mass [283]). It would be correct to say that there are a number of experimental limits to the mass of the photon. These limits have become more and more precise [284, 165]. The best limit so far is given by

$$m_\gamma < 10^{-57} gm \tag{4.28}$$

that is, the photon mass would be very small indeed!
It may also be mentioned that there has been a more radical view that the photon itself is superfluous [285–288].
It is interesting that we can pursue the gravito thermodynamic link with Electromagnetism further. In fact if we start with the Langevin equation in a viscous medium as we did in Chapter 2 [18, 138] then as the viscosity becomes vanishingly small, it turns out that the Brownian particle moves according to Newton's first law, that is with a constant velocity. Moreover this constant velocity is given by (Cf.refs. [18, 138]), for any mass m,

$$\langle v^2 \rangle = \frac{kT}{m} \tag{4.29}$$

We would like to study the case where $m \to 0$. Then so too should T for a meaningful limit. More realistically, let us consider (4.29) with minimal values of T and m, in the real world. We consider in (4.18) the entire Universe so that the mass M is $\sim 10^{55} gms$. The justification for this is that the Universe mimics a black hole, as shown in detail several years ago by the author and as noted earlier. This can be seen in a simple way by the fact that as we saw in the last Chapter, the size of the Universe is given by the Schwarzchild radius:

$$R \approx \frac{2GM}{c^2}$$

Further the time taken by light to reach the boundary at a distance R from a given point is the same as for a black hole of the mass of the Universe (Cf.refs.[130] and [44] for further details). Substitution in (4.18) gives

$$T = \frac{10^4}{10^{32}} \sim 10^{-28}K \quad (4.30)$$

We next consider in (4.29), m to be the smallest possible mass encountered earlier, viz.,

$$m \sim 10^{-65} gms \quad (4.31)$$

Equation (4.31) has been obtained from different points of view, namely the Planck scale underpinning for the Universe and from thermodynamics as seen. Substitution of (4.30) and (4.31) in (4.29) gives

$$\langle v^2 \rangle = \frac{kT}{m} = \frac{10^{-16} \times 10^{-28}}{10^{-65}} = 10^{21}, i.e.$$

$$v = c\,(cm/sec) \quad (4.32)$$

We can see from (4.29) and (4.32) that the velocity c is the velocity of light! So m in (4.31) indeed represents the mass of the photon. We can derive this velocity of light (4.32) in a heuristic way from a different angle. We have already seen that the ZPF causes a harmonic motion. Let us assume that the particle has a small charge e, just to couple it to the ZPF. The equation of motion is now given by (Cf.[289])

$$\ddot{x} + \omega^2 J = (e/m)E_x^0$$

along the x axis, where, suppressing the polarization states for the moment, the random field $\vec{E}$ is given by

$$\vec{E} \propto \int d^3\vec{K} exp[-\imath(\omega t - K \cdot \vec{r})]\vec{a}_K + c \cdot c \cdot$$

where, owing to the randomness in phase, their averages vanish. What this means is that, finally,

$$L^2 = \langle x^2 \rangle = \hbar/(2m\omega) \,\text{and}\, \langle \dot{x} \rangle^2 = (\hbar/2m)\,\omega = v^2 \quad (4.33)$$

In (4.33) above, the frequency is given by,

$$\omega = mc^2/\hbar$$

Using this in the above, with $m_\gamma \sim 10^{-65} gm$ we get, for L, the radius of the Universe and for v, the velocity of light.
The question is, for how long such a particle with vanishingly small inertial mass can maintain the constant velocity c, that is the velocity of light. In fact the energy uncertainty mc^2 is associated with the lifetime $\sim \hbar/mc^2$, the Compton time. The Compton time for a particle with the mass given by (4.31) is, as can be easily checked $10^{17} secs$, which is the age of the Universe! We should recover the same result from (4.21) and indeed we do!
We have argued that the photon has a small mass given by (4.31),

$$m_\gamma = m = 10^{-65} gm$$

This conclusion follows from the Planck scale underpinning model and is compatible with special relativity [18, 19, 277, 16] and also from thermodynamics. In any case (4.31) is compatible with the limit (4.28).
It is interesting to note that the mass of the photon given in (4.31) has some experimental support. This mass would imply a dispersive effect in High Energy Cosmic Rays which we receive from deep outer space−−and it appears that we may already be observing such effects as we will see later [19, 290].
This apart another experimental evidence for the photon mass (4.31) comes from laboratory diffraction experiments [291]. In these experiments, it turns out that the vacuum is a dissipative medium, which has been our theoretical point of view as seen in Chapters 2 and 3.
Finally it may be mentioned that this photon mass would imply that the Coulomb potential becomes a short range Yukawa potential with a range $\sim 10^{28} cms$, that is the radius of the Universe itself. Such a Yukawa potential would lead to a small shift in the hyperfine energy levels as we will see shortly [292].

4.5 Further Theoretical Support

We now give a further theoretical justification for the above. We first observe that as is well known [160], Maxwell's equations can be written in the following form

$$\Psi = \vec{E} + \imath \vec{B}, \tag{4.34}$$

$$\vec{\nabla} \times \mathbf{\Psi} = \imath \dot{\mathbf{\Psi}} \tag{4.35}$$

$$\vec{\nabla} \cdot \mathbf{\Psi} = 0 \tag{4.36}$$

Equations (4.34) to (4.36) will be useful in the sequel.

We next observe that Maxwell's equations have been deduced in a fashion very similar to the Dirac equation, from first principles [293]. In this deduction, we use the usual energy momentum relation for the photon

$$E^2 - p^2c^2 = 0$$

and introduce matrices given in

$$S_x = \begin{pmatrix} 0 & 0 & 0 \\ 0 & 0 & -\imath \\ 0 & \imath & 0 \end{pmatrix}, S_y = \begin{pmatrix} 0 & 0 & \imath \\ 0 & 0 & 0 \\ -\imath & 0 & 0 \end{pmatrix},$$

$$S_z = \begin{pmatrix} 0 & -\imath & 0 \\ \imath & 0 & 0 \\ 0 & 0 & 0 \end{pmatrix}, I^{(3)} = \begin{pmatrix} 1 & 0 & 0 \\ 0 & 1 & 0 \\ 0 & 0 & 1 \end{pmatrix}, \tag{4.37}$$

from which we get

$$\left(\frac{E^2}{c^2} - \mathbf{p}^2\right)\mathbf{\Psi} = \left(\frac{E}{c}I^{(3)} + \mathbf{p} \cdot \mathbf{S}\right)\mathbf{\Psi}$$

$$- \begin{pmatrix} p_x \\ p_y \\ p_z \end{pmatrix} (\mathbf{p} \cdot \mathbf{\Psi}) = 0, \tag{4.38}$$

where Ψ is a three component wave function and in general bold letters denote vector quantities.

Equation (4.38) implies

$$\left(\frac{E}{c}I^{(3)} + \mathbf{p} \cdot \mathbf{S}\right)\mathbf{\Psi} = 0, \tag{4.39}$$

$$\mathbf{p} \cdot \mathbf{\Psi} = 0, \tag{4.40}$$

where S is given in (4.37). There is also an equation for $\mathbf{\Psi}^*$ namely

$$\left(\frac{E}{c}I^{(3)} - \mathbf{p} \cdot \mathbf{S}\right)\mathbf{\Psi}^* = 0, \tag{4.41}$$

$$\mathbf{p} \cdot \mathbf{\Psi}^* = 0, \tag{4.42}$$

It is then easy to verify (Cf.ref.[293]) that with the substitution of the usual Quantum Mechanical energy momentum operators, we recover equations (4.34) to (4.36) for $\boldsymbol{\Psi}$ and its complex conjugate.
Recently a similar analysis has lead to the same conclusion. In fact it has been shown that under a Lorentz boost [294–296],

$$\begin{pmatrix} \Psi' \\ \Psi^{*'} \end{pmatrix} = \begin{pmatrix} 1 - \frac{(\mathbf{S}\cdot\mathbf{p})}{mc} + \frac{(\mathbf{S}\cdot\mathbf{p})^2}{m(E+mc^2)} & 0 \\ 0 & 1 + \frac{(\mathbf{S}\cdot\mathbf{p})}{mc} + \frac{(\mathbf{S}\cdot\mathbf{p})^2}{m(E+mc^2)} \end{pmatrix} \begin{pmatrix} \Psi \\ \Psi^* \end{pmatrix} \qquad (4.43)$$

We would like to point out that equations (4.35), (4.36), (4.39) to (4.43) display the symmetry

$$\mathbf{p} \to -\mathbf{p} \quad , \Psi \to \Psi^*$$

We now invoke the Weinberg-Tucker-Hammer formalism (Cf.[296]) which gives, for a Lorentz boost, the equations

$$\phi'_R = \left\{ 1 + \frac{\mathbf{S}\cdot\mathbf{p}}{m} + (\mathbf{S}\cdot\mathbf{p})^2 m(E+m) \right\} \phi_R, \qquad (4.44)$$

$$\phi'_L = \left\{ 1 - \frac{\mathbf{S}\cdot\mathbf{p}}{m} + (\mathbf{S}\cdot\mathbf{p})^2 m(E+m) \right\} \phi_L, \qquad (4.45)$$

where the subscripts R and L refer to the states of opposite helicity, that is left and right polarised light in our case.
We observe that equations (4.43) and (4.44)-(4.45) are identical, but there is a curious feature in both of these, that is that the photon of electromagnetism is now seen to have a mass m.

4.6 Remarks

It is interesting to note that it has been demonstrated that the mass of the photon is incompatible with the magnetic monopole [295]. Indeed the author himself has presented different arguments to the effect that there are no magnetic monopoles [297]. We will encounter that argument in the last Chapter. It may be mentioned that Dirac the originator of the idea of the magnetic monopole, himself expressed his pessimism about the existence of the magnetic monopole as long back as 1981, during the fiftieth year of the monopole seminar [298].
Returning to the mass of the photon, it can be argued that this is a result of the non commutativity of spacetime at a micro scale. We observe that a photon mass would imply the equation

$$\partial^\mu F_{\mu\nu} = -m^2 A_\nu, \qquad (4.46)$$

where we have the usual equations of electromagnetism

$$\partial^\mu A_\mu = 0, \tag{4.47}$$

$$F_{\mu\nu} = \partial_\mu A_\nu - \partial_\nu A_\mu \tag{4.48}$$

We note that from (4.48) we get

$$\partial^\mu F_{\mu\nu} = DA_\nu - \partial^\mu \partial_\nu A_\mu \tag{4.49}$$

where D denotes the D'Alembertian. In view of (4.47), the second term on the right side of (4.49) would vanish, provided the derivates commute. In this case we would return to the usual Maxwell equations. However in the non commutative case this extra term is

$$p^2 A_\mu \sim m^2 A_\mu$$

remembering that we are in units $c = 1 = \hbar$.
Thus because of the non commutativity we get (4.46) instead of the usual Maxwell equation.
The question of non commutativity and mass generation in the context of gauge theory has been studied by the author elsewhere (Cf.[299, 300, 16]).
We give another but this time heuristic estimate for the photon mass. Let us consider the "large scale" part of the Uncertainty Principle, encountered in the previous Chapter, viz., Equation (3.79):

$$\Delta x = l^2 \cdot \frac{\Delta p}{\hbar}$$

If we take the whole Universe into consideration, we have

$$\Delta x = l$$

the radius of the Universe. So if the photon has a mass m_γ, then we have

$$m_\gamma c \approx \hbar / R$$

This gives back the photon mass. Interestingly, the same conclusion can be obtained by using the usual Uncertainty relation. This means that the mass m_γ emphasizes that the photon's position is indeterminate within the Universe and likewise for its energy having the life time of the Universe.
In any case the above points to the fact that there would be no massless particles in nature. The point is, that in an idealized situation in which the radius $R \to \infty$, (and, as we saw earlier, the number of particles $N \to \infty$), the mass $m_\gamma \to 0$.
We briefly consider some of the consequences of the photon mass and also look for experimental verification, apart from consistency with theory.

It may be reiterated that the mass for the photon has been proposed in the past, though from phenomenological considerations [301, 302]. Indeed it is remarkable that exactly the above mass was indicated from experimental observation (Cf.ref.[291] and references there in), and has been attributed as noted to a vacuum induced dissipative mechanism.

With a non zero photon mass we would have, for radiation

$$E = h\nu = m_\nu c^2 [1 - v_\gamma^2/c^2]^{-1/2} \tag{4.50}$$

From (4.50) one would have a dispersive group velocity for waves of frequency ν given by (Cf. also ref.[291])

$$v_\gamma = c \left[1 - \frac{m_\gamma^2 c^4}{h^2 \nu^2} \right]^{1/2} \tag{4.51}$$

We would like to point out that (4.51) indicates that higher frequency radiation has a velocity greater than lower frequency radiation. This is a very subtle and minute effect and is best tested in for example, the observation of high energy gamma rays, which we receive from deep outer space. It is quite remarkable that we may already have witnessed this effect– higher frequency components of gamma rays in cosmic rays do indeed seem to reach earlier than their lower frequency counterparts [290]. The GLAST satellite of NASA to be launched in 2007 may be able to throw more light on these high energy Gamma rays. We will return to these considerations in Chapter 5.

Another result of the non zero mass of the photon is that in addition to the two traverse polarizations of light, there will be a longitudinal component also (Cf. for example ref.[303]).

This apart, a finite photon mass would imply a slight modification of the Coulomb interaction, which would go over into a Yukawwa type potential, this given by, (in natural units $\hbar = 1 = c$.)

$$V(r) = e^{-\mu r}/r \tag{4.52}$$

where μ is the mass in these units. As can be seen from (4.52), the potential V has a finite range. However this range is $\sim \frac{1}{\mu}$ which is $\sim 10^{28} cms$, as can be easily calculated. This range is in fact the radius of the Universe! Nevertheless this cut off does imply that there will not be any infra red divergences [165].

The range of the Yukawa type Coulomb potential (4.52) being of the order of the radius of the Universe, we would expect that the modification would be miniscule. Nevertheless from a strictly mathematical point of view, the photon mass converts the otherwise long range Coulomb potential into a short range Yukawa potential with the consequence that several otherwise strictly divergent integrals become convergent. It then becomes possible to use techniques suitable for short range potentials (Cf. for example [303, 304]). The rather laborious modifications required for handling the Coulomb potential in scattering theory (Cf. for example [305, 306]) can be eased. For instance there is an interesting recurrence relation for the large l phase shifts of the Yukawa potential [307] viz.,

$$\delta_{l+1}/\delta_l \approx 1 - (\mu/K)$$

valid for a large range of energies K. This relation can now be used, though as μ is very small, it shows that the convergence of the phase shifts with respect to l is extremely slow.
Let us introduce the Yukawa potential,

$$V(r) = \alpha e^{-\mu r}/r \tag{4.53}$$

into the Dirac equation instead of the usual Coulomb potential [308]. In (4.53), μ is proportional to the mass of the photon $\sim 10^{-65} gms$ and is therefore a very small quantity. We introduce (4.53) into the stationary Dirac equation to get

$$[c\vec{\alpha} \cdot \vec{p} + \beta m_0 c^2 - (E - V(r))]\psi(r) = 0 \tag{4.54}$$

From (4.54) and (4.53), we can immediately see that roughly the effect of the photon mass is to shift the energy levels by a miniscule amount.
We further introduce the notation

$$Q = 2\lambda r, \quad \text{where}\, \lambda = \frac{\sqrt{m_0^2 c^4 - E^2}}{hc}, \tag{4.55}$$

After the standard substitutions (Cf.ref.[308]) we finally obtain

$$\frac{d\Phi_1}{dQ} = \left(1 - \frac{\alpha E}{hc\lambda Q}\right)\Phi_1 - \left(\frac{\chi}{Q} + \frac{\alpha m_0 c^2}{hc\lambda Q}\right)\Phi_2,$$

$$\frac{d\Phi_2}{dQ} = \left(-\frac{\chi}{Q} + \frac{\alpha m_0 c^2}{hc\lambda Q}\right)\Phi_1 + \left(\frac{\alpha E}{hc\lambda Q}\right)\Phi_2 \tag{4.56}$$

The substitutions

$$\Phi_1(Q) = Q^\gamma \sum_{m=0}^{\infty} \alpha_m Q^m,$$

$$\Phi_2(Q) = Q^\gamma \sum_{m=0}^{\infty} \beta_m Q^m. \tag{4.57}$$

in (4.35) leads to

$$\alpha_m(m+\gamma) = \alpha_{m-1} - \left(\frac{\alpha E}{hc\lambda}\right)\alpha_m - \left(\chi + \frac{\alpha m_0 c^2}{hc\lambda}\right)\beta_m,$$

$$\beta_m(m+\gamma) = \left(-\chi + \frac{\alpha m_0 c^2}{hc\lambda}\right)\alpha_m + \left(\frac{\alpha E}{hc\lambda}\right)\beta_m. \tag{4.58}$$

As is well known γ in (4.57) is given by

$$\gamma = \pm\sqrt{\chi^2 - \alpha^2}, \tag{4.59}$$

where λ is given by (4.55).
At this stage we remark that the usual method adopted for the Coulomb potential is no longer valid—–mathematically, Sommerfeld's polynomial method becomes very complicated and even does not work for a general potential: We have to depart from the usual procedure for the Coulomb potential in view of the Yukawa potential (4.53). Nevertheless, it is possible to get an idea of the solution by a slight modification. This time we have from (4.58), instead the equations

$$\alpha_m(m+\gamma) = \alpha_{m-1} - \frac{\alpha E}{\hbar c\lambda}\alpha_m + \frac{\alpha E\mu}{\hbar c\lambda}\alpha_{m-1}$$

$$(-\chi\beta_m) - \frac{\alpha m_o c^2}{\hbar c\lambda}\beta_m + \frac{\mu\alpha m_0 c^2}{\hbar c\lambda}\beta_{m-1} \tag{4.60}$$

and

$$(m+\gamma)\beta_m = -\left(\chi + \frac{\alpha m_0 c^2}{\hbar c\lambda}\right)\alpha_m - \frac{\mu\alpha m_0 c^2}{\hbar c\lambda}\alpha_{m-1}$$

$$+\frac{\alpha E}{\hbar c\lambda}\beta_m - \frac{\mu\alpha E}{\hbar c\lambda}\beta_{m-1} \tag{4.61}$$

After some algebra on (4.60) and (4.61) we obtain

$$P\alpha_m + Q\beta_m = R\alpha_{m-1} \tag{4.62}$$

$$S\alpha_m + T\beta_m = U\beta_{m-1} \tag{4.63}$$

where P, Q, S, T can be easily characterized, in the derivation of which we will neglect μ^2 and higher orders.
If

$$\alpha_m/\beta_m \equiv p_m$$

then we have from (4.62) and (4.63),

$$Sp_m + T = \frac{U(QS - PT)}{(RSp_{m-1} - PU)}$$

We note that the asymptotic form of the series in (4.57) will not differ much from the Coulomb case and so we need to truncate these series. For the truncation of the series we require

$$QS = PT$$

This gives

$$\left\{1 + \frac{\chi\hbar c\lambda}{\alpha m_0 c^2}\right\}\left[E\left(\chi + \frac{\alpha m_0 c^2}{\hbar c\lambda}\right) + m_0 c^2\left(m + \gamma - \frac{\alpha E}{\hbar c\lambda}\right)\right]$$

$$= \left(m + \gamma - \frac{\alpha E}{\hbar c\lambda}\right)\frac{\hbar c\lambda}{\alpha m_0 c^2}\left[E\left(m + \gamma + \frac{\alpha E}{\hbar c\lambda}\right) + m_0 c^2\left(\chi + \frac{\alpha m_0 c^2}{\hbar c\lambda}\right)\right]$$

Further simplification yields

$$(m + \gamma)^2 + \left\{\frac{\chi\hbar c\lambda + \alpha m_0 c^2}{\hbar c\lambda}\right\}^2 + \frac{\alpha^2 E^2}{\hbar^2 c^2 \lambda^2}$$

where γ is given by (4.59). Finally we get, in this approximation,

$$E^2 = m_0^2 c^4\left[1 - \frac{2\alpha^2}{\alpha^2 + (m + \gamma)^2 - \chi^2}\right] + AO(\frac{1}{m^2}) \tag{4.64}$$

In (4.64) A is a small quantity

$$A \sim m_0^2 c^4 - E^2$$

The second term in (4.64) is a small shift from the usual Coulombic energy levels. In (4.64) m is a positive integer and this immediately provides a comparison with known fine structure energy levels. To see this further let us consider large values of m. (4.64) then becomes

$$E = m_0 c^2\left[1 - \frac{\alpha^2}{m_1^2}\right] \tag{4.65}$$

while the usual levels are given by

$$E = m_0 c^2\left[1 - \frac{\alpha^2}{2m_2^2}\right] \tag{4.66}$$

We can see from (4.65) and (4.66) that the photon mass reproduces all the energy levels of the Coulomb potential but interestingly (4.65) shows that there are new energy levels because m_1^2 in the new formula can be odd or even but $2m_2^2$ in the old formula is even. However all the old energy levels are reproduced whenever $m_1^2 = 2m_2^2$. If $\mu = 0$, then as can be seen from (4.54), we get back the Coulomb problem. In any case, the above calculation was suggestive and more to indicate how the problem changes.

It must be remembered that all these effects are small and consistent with the size of the Universe. Nevertheless there are experimental tests, in addition to those mentioned above, which are doable. It is well known that for a massive vector field interacting with a magnetic dipole of moment **M**, for example the earth itself, we would have with the usual notation (Cf.ref.[165])

$$\mathbf{A}(x) = \frac{\imath}{2} \int \frac{d^3k}{(2\pi)^3} \mathbf{M} \times \mathbf{k} \frac{e^{\imath k,x}}{\mathbf{k}^2 + \mu^2} = -\mathbf{M} \times \nabla \left(\frac{e^{-\mu r}}{8\pi r} \right)$$

$$\mathbf{B} = \frac{e^{-\mu r}}{8\pi r^3} |\mathbf{M}| \left\{ \left[\hat{r}(\hat{r} \cdot \hat{z}) - \frac{1}{3}\hat{z} \right] (\mu^2 r^2 + 3\mu r + 3) - \frac{2}{3}\hat{z}\mu^2 r^2 \right\} \qquad (4.67)$$

Considerations like this have yielded as noted in the past an upper limit for the photon mass, for instance $10^{-48} gms$ and $10^{-57} gms$. Nevertheless (4.67) can be used for a precise determination of the photon mass. It may be reiterated here that contrary to popular belief, there is no experimental evidence to indicate that the photon mass is zero! (Cf. discussion in ref. [301]). We will return to this in a later Chapter.

Finally, it is interesting to observe that the above value for the photon mass was also obtained by Terazawa [309], using the Dirac Large Number Hypothesis, something which is in fact a consequence of our Planck oscillator approach.

4.7 The Mass Spectrum

We have deduced the minimum and maximum masses as also the Planck mass in earlier sections. We also indicated how it is possible to obtain the mass scale of elementary particles, using the Compton length. We now refine the argument. One of the problems that has eluded a solution is precisely that of a mass spectrum for elementary particles as noted in Chapter 1. In other words, why should there be such a plethora of particles, and why should they have such and such masses? Is there any formula, based on dynamics, which would generate all these known masses? Such a formula would be intimately tied up with inter quark interactions which as mentioned are tied up to the ZPF. We will now use the inter quark or QCD potential to deduce such a formula, which as will be seen, surprisingly covers all known elementary particles besides predicting any number of others. The well known QCD potential we encountered in Chapter 1 is given by [30, 130]

$$U(r) = -\frac{\alpha}{r} + \beta r \tag{4.68}$$

where in units $\hbar = c = 1, \alpha \sim 1$. The first term in (4.68) represents the Coulombic part while the second term represents the confining part of the potential. The potential in (4.68) explains two well known features viz., quark confinement and asymptotic freedom. We will see in a later Chapter how (4.68) follows from our theory.
Let us consider the pion made up of two quarks along with a third quark, one at the centre and two at the ends of an interval of the order of the Compton wavelength, r. Then the central particle experiences the force

$$\frac{\alpha}{(\frac{x}{2}+r)^2} - \frac{\alpha}{(\frac{x}{2}-r)^2} \approx \frac{-2\alpha x}{r^3} \tag{4.69}$$

where x is the small displacement from the mean position. Equation (4.69) gives rise to the Harmonic oscillator potential, and the whole configuration resembles the tri-atomic molecule. This is pleasingly similar to the oscillator scenario we encountered earlier wherein the pions themselves were made up of Planck scale oscillators.
Before proceeding we can make a quick check on (4.69). We use the fact that the frequency is given by

$$\omega = \left(\frac{\alpha^2}{m_\pi r^3}\right)^{\frac{1}{2}} = \frac{\alpha}{(m_\pi r^3)^{\frac{1}{2}}}$$

whence the mass of the pion m_π is given by

$$(\hbar\omega \equiv)\omega = m_\pi \tag{4.70}$$

Remembering that $r = 1/m_\pi$, use of (4.70) gives $\alpha = 1$, which of course is correct.
To proceed, the energy levels of the Harmonic oscillator are now given by,

$$\left(n + \frac{1}{2}\right) m_\pi,$$

If there are N such oscillators, then over the various modes the energy of the particle is given by

$$E = \sum_{r=1}^{3N} \left(n_r + \frac{1}{2}\right) \hbar\omega = m\left(n + \frac{1}{2}\right) \hbar\omega$$

m and n being positive integers. The mass of the particle P is now given by

$$m_P = m\left(n + \frac{1}{2}\right)m_\pi \tag{4.71}$$

where m_P now is the mass of the corresponding elementary particle. The formula (4.71) gives the mass of all known elementary particles with an error of less than one percent for sixty three percent of the particles, less than two percent for ninety three percent of the particles, and less than three percent for all particles with the lone exception of $\omega(782)$, in which case the error is 3.6%. The known elementary particles for which the formula (4.71) is valid include the recently discovered $Ds(2317)$ and the $1.5GeV$ Pentaquark, all be it still debated, discovered after the above formula was deduced. All the values are displayed in the Tables.
In fact it is surprising that there is such a good fit for all the particles [310] considering that only bare details of the interaction have been taken into consideration. Once other details are included, the agreement could be even better. It may also be mentioned that a similar approach, but using the proton as the base particle had lead to interesting, but not such comprehensive results [311–313, 42, 314]. Furthermore, instead of starting with quarks in (4.68) or (4.69), we could have very well started with electrons and positrons. Indeed, this would be in the spirit of the fact that quarks are electrons at a smaller scale [130].

4.8 Further Remarks

We have already seen in the previous Chapter, that the indeterminable part in General Relativity due to the mass distribution can be treated as a fluctuation in the curvature. This leads to Quantum theory from the Classical theory−−specifically, it characterizes the mysterious Quantum Potential that appears in the transition from classical to Quantum theory. We have also seen how Quantum theory can be considered to be a result of large scale cosmic fluctuations, very much in the spirit of Chapter 2.
Another way of looking at this is that the large scale fluctuation in gravitational energy of elementary particles is the same as the energy of fluctuation in the ZPF, given in (4.1). That is,

$$\frac{G\sqrt{N}m^2}{R} = \hbar c/R \tag{4.72}$$

Equation (4.72) gives us the value of $\hbar c$ in terms of G, N and m. Equivalently we could also get the residual energy $\sim 10^{-33}eV$ alluded to from the left side of (4.72), this being also the mass of the photon in our theory.
In any case, the point we have been putting forward is, that the Universe is "thermodynamic" in the sense that it is governed by large scale fluctuations, rather than being, as Prigogine noted (quoted in Chapter 2), a "mechanistic" process.
We finally make the following remark. We have already seen that a particle in the ZPF modeled by a Random electromagnetic field, leads to equation (4.33), viz.,

$$L^2 = \langle x^2 \rangle = \hbar/2m\omega,$$

$$\langle \dot{x} \rangle = (\hbar/2m)\omega,$$

where the frequency is given $\omega = mc^2/\hbar$. In the equation leading to (4.33), we could also include a term which gives the third derivative of x, this corresponding to the Schott term we saw in Chapter 1. However the contribution of this term, which was introduced for energy conservation to compensate the radiation loss of an accelerated charge [6] in the classical dectron electron theory, is of the order of the Compton wavelength [14] and does not effect the conclusion. The above show that there is oscillation of the particle within the Compton wavelength L, with velocity c. This models the well known zitterbewegung of Dirac. However, what all this means is that via the Compton length L, we get the inertial mass of the particle, which is now seen to be due to the energy of this oscillation. All this very pleasingly in accordance with our earlier conclusions also (Cf.[129]).

Particle and Mass	Mass from Formula	Error %	(l, n)
$p(938) \pm 0.00008$	959	$-2.23881,$	$(2,3)$
$n(939)939.56533 \pm 0.00004$	959	$-2.12993,$	$(2,3)$
$P_{11}N(1440)(1420-1470)$	1438.5	$(0.138889,)0$	$(1,10)$
$D_{13}N(1520)(1515-1525)$	1507	$(0.855263,)$	$(2,5)$
$S_{11}N(1535)(1525-1545)$	1507	1.9442	$(2,5)$
$S_{11}N(1650)(1645-1670)$	1644	$(0.363636,)0$	$(8,1)$
$D_{15}N(1675)(1670-1680)$	1644	1.85075,	$(8,1)$
$F_{15}N(1680)(1680-1690)$	1644	2.14286,	$(8,1)$
$D_{13}N(1700)(1650-1750)$	1712.5	$(-0.705882,)0$	$(1,12)$
$P_{11}N(1710)(1680-1740)$	1712.5	$(-0.116959,)0$	$(1,12)$
$P_{13}N(1720)(1700-1750)$	1712.5	$(0.465116,)0$	$(1,12)$
$P_{13}N(1900)$	1918	$-0.947368,$	$(4,3)$
$F_{17}N(1990)$	1986.5	0.201005,	$(1,14)$
$F-15N(2000)$	1986.5	0.7,	$(1,14)$
$D_{13}N(2080)$	2055	1.20192,	$(2,7)$
$S_{11}N(2090)$	2123.5	$-1.57895,$	$(1,15)$
$P_{11}N(2100)$	2123.5	$(-1.09524,)$	$(1,15)$
$G_{17}N(2190)(2100-2200)$	2123.5	$(3.05936,)0$	$(1,15)$
$D_{15}N(2200)$	2260.5	$-2.72727,$	$(3,5)$
$H_{19}N(2220)(2200-2300)$	2260.5	$(-1.8018,)0$	$(3,5)$
$G_{19}N(2250)(2200-2350)$	2260.5	$(-0.444444,)0$	$(3,5)$
$I_{1;11}N(2600)(2550-2750)$	2603	$(-0.115385,)0$	$(2,9)$
$K_{1;13}N(2700)$	2671.5	1.05556	$(1,19)$
$P_{33}\Delta(1232)(1231-1233)$	1233	$(-0.0811688,)0$	$(2,4)$
$P_{33}\Delta(1600)(1550-1700)$	1575.5	$(1.5625,)0$	$(1,11)$
$S_{31}\Delta(1620)(1600-1660)$	1644	$(-1.46148,)0$	$(8,1)$
$D33\Delta(1700)(1670-1750)$	1712	$(-0.705882,)0$	$(1,12)$
$P31\Delta(1750)$	1781	$-1.77143,$	$(2,6)$
$S_{31}\Delta(1900)$	1918	$-0.947368,$	$(4,3)$
$F35\Delta(1905)(1865-1915)$	1918	$(-0.682415,)0$	$(4,3)$
$P31\Delta(1910)(1870-1920)$	1918	$(-0.418848,)0$	$(4,3)$
$P33\Delta(1920)(1900-1970)$	1918	$(0.104167,)0$	$(4,3)$
$D35\Delta(1930)(1900-2020)$	1918	$(0.621762,)0$	$(4,3)$
$D33\Delta(1940)$	1918	1.13402,	$(4,3)$

Particle and Mass	Mass from Formula	Error %	(l, n)
$F37\Delta(1950)(1915-1950)$	1918	1.64103,	(4, 3)
$F35\Delta(2000)$	1986	0.7,	(1, 14)
$S_{31}\Delta(2150)$	2123.5	1.25581,	(1, 15)
$G_{37}\Delta(2200)$	2260	−2.72727,	(1, 16)
$H_{39}\Delta(2300)$	2329	−1.26087,	(2, 8)
$D_{35}\Delta(2350)$	2329	0.893617,	(2, 8)
$F_{37}\Delta(2390)$	2397.5	−0.292887,	(1, 17)
$G_{39}\Delta(2400)$	2397.5	0.125,	(1, 17)
$H_{3;11}\Delta(2420)(2300-2500)$	2397.5	(0.950413,)0	(1, 17)
$I_{3;13}\Delta(2750)$	2740	0.363636,	(8, 2)
$K_{3;15}\Delta(2950)$	2945.5	0.152542,	(1, 21)
$\Lambda(1115) \pm 0.006$	1096	1.7000,	(16, 0)
$P_{01}\Lambda(1600)(1560-1700)$	1575.5	1.53125,	(1, 11)
$S_{01}\Lambda(1405) \pm 4$	1438.5	−2.3130,	(1, 10)
$D_{03}\Lambda(1520) \pm 1.0$	1507	0.855263,	(2, 5)
$P01\Lambda(1600)(1560-1700)$	1575.5	(1.5625,)0	(1, 12)
$S01\Lambda(1670)(1660-1680)$	1644	1.55689,	(8, 1)
$D03\Lambda(1690)(1685-1695)$	1712.5	−1.30178,	(1.12)
$S01\Lambda(1800)(1720-1850)$	1781	(1, 05556,)0	(2, 6)
$P01\Lambda(1810)(1750-1850)$	1781	(1.60221,)0	(2, 6)
$F05\Lambda(1820)(1815-1825)$	1849.5	(2.14286,)	(1, 13)
$D05\Lambda(1830)(1810-1830)$	1849.5	−1.03825,	(1, 13)
$P03\Lambda(1890)(1850-1910)$	1918	−1.48148,	(4, 3)
$*\Lambda(2000)$	1986.5	0.7,	(1, 14)
$F07\Lambda(2020)$	2055	−1.73267,	(2, 7)
$G07\Lambda(2100)(2090-2110)$	2123.5	−1.09524,	(1, 15)
$F05\Lambda(2110)(2090-2140)$	2123.5	(−0.616114,)0	(1, 15)
$D03\Lambda(2325)$	2329	−0.172043,	(2, 8)
$H09\Lambda(2350)(2340-2370)$	2329	0.893617,	(2, 8)
$\Lambda(2585)$	2603	0.309478,	(2, 9)
$P11\Sigma+(1189.37) \pm 0.07$	1164.5	2.10261,	(1, 8)
$P11\Sigma0(1192.642) \pm 0.024$	1164.5	2.34899,	(1, 8)
$\Sigma-(1197.440) \pm 0.030$	1164.5	2.75689,	(1, 8)
$P13\Sigma(1385) \pm 0.4$	1370	(0.108),	(4, 2)
$\Sigma(1480)$	1438.5	2.83784,	(1, 10)
$\Sigma(1560)$	1575.5	−0.961538,	(1, 11)

Particle and Mass	Mass from Formula	Error %	(l, n)
$D13\Sigma(1580)$	1575.5	0.316456,	$(1,11)$
$S11\Sigma(1620)$	1644	−1.48148,	$(8,1)$
$P11\Sigma(1660)(1630-1690)$	1644	(0.963855,)0	$(8,1)$
$D13\Sigma(1670)(1665-1685)$	1644	1.55689,	$(8,1)$
$\Sigma(1690)$	1712.5	−1.30178,	$(1,12)$
$S11\Sigma(1750)(1730-1800)$	1781	(−1.77143,)0	$(2,6)$
$P11\Sigma(1770)$	1781	−0.621469,	$(2,6)$
$D15\Sigma(1775)(1770-1780)$	1781	(−0.338028,)0	$(2,6)$
$P13\Sigma(1840)$	1849.5	−0.48913,	$(1,13)$
$P11\Sigma(1880)$	1849.5	1.64894,	$(1,13)$
$F15\Sigma(1915)(1900-1935)$	1918	(−0.156658,)0	$(4,3)$
$D13\Sigma(1940)(1900-1950)$	1918	(1.13402,)0	$(4,3)$
$S11\Sigma(2000)$	1986.5	0.7,	$(1,14)$
$F17\Sigma(2030)(2025-2040)$	2055	−1.23153,	$(2,7)$
$F15\Sigma(2070)$	2055	0.724638,	$(2,7)$
$P13\Sigma(2080)$	2055	1.20192,	$(2,7)$
$G17\Sigma(2100)$	2123	−1.09524,	$(1,15)$
$\Sigma(2250)(2210-2280)$	2260	(−0.444444,)0	$(3,5)$
$\Sigma(2455)$	2466	−0.448065,	$(4,4)$
$\Sigma(2620)$	2603	0.648855,	$(2,9)$
$\Sigma(3000)$	3014	−0.466667,	$(4,5)$
$\Sigma(3170)$	3151	0.599369,	$(2,11)$
$P11\Xi0, \Xi-(1314.83)\pm 0.20$	1301.5	1.01156,	$(1,9)$
$\Xi(1321)\pm 0.13$	1301.5	1.47615,	$(1,9)$
$P13\Xi(1530)\pm 0.32$	1507	1.50327,	$(2,5)$
$\Xi(1620)$	1644	−1.48148,	$(8,1)$
$\Xi(1690)\pm 10$	1712.5	−1.30178,	$(1,12)$
$D13\Xi(1820)\pm 5$	1849.5	−1.59341,	$(1,13)$
$\Xi(1950)\pm 15$	1918	1.64103,	$(4,3)$
$\Xi(2030)\pm 5$	2055	−1.23153,	$(2,7)$
$\Xi(2120)$	2123.5	−0.141509,	$(1,15)$
$\Xi(2250)$	2260.5	−0.444444,	$(1,16)$
$\Xi(2370)$	2397.5	−1.13924,	$(1,17)$
$\Xi(2500)$	2534.5	−1.36,	$(1,18)$

Particle and Mass	Mass from Formula	Error %	(l, n)
$\Omega-(1672)\pm 0.29$	1644	1.67464,	$(8,1)$
$\Omega-(2250)\pm 9$	2260.5	$(-0.444444,)0$	$(1,16)$
$\Omega-(2380)$	2397.5	$-0.714286,$	$(1,17)$
$\Omega-(2470)$	2466	0.161943,	$(4,4)$
$\Lambda c+2286.46\pm 0.14$	2260.5	1.09409,	(1.16)
$\Lambda c+(2593)2595.4\pm 0.6$	2603	$-0.385654,$	$(2,9)$
$\Lambda c+(2625)2628.1\pm 0.6$	2603	0.838095,	$(2,9)$
$\Lambda c+(2765)$	2740	0.904159,	$(8,2)$
$\Lambda c+(2880)$	2877	0.104167,	$(2,10)$
$\Sigma c(2455)2454.02\pm 0.18$	2466	$-0.448065,$	$(4,4)$
$\Sigma c(2520)2518.4\pm 0.6$	2534.5	$-0.555556,$	$(1,18)$
$\Xi c+(2466)2467.9\pm 0.4$	2466	0,	$(4,4)$
$\Xi c0(2471)2471.0\pm 0.4$	2466	0.202347,	$(4,4)$
$\Xi c+(2574)2575.7\pm 3.1$	2603	$(1.12665,)0$	$(2,9)$
$\Xi c0(2578)\pm 2.9$	2603	$(0.96974,)0$	$(2,9)$
$\Xi c(2645)2646.6\pm 1.4$	2671.5	$-0.982987,$	$(1,19)$
$\Xi c(2790)2789.2\pm 3.2$	2808.5	$-0.645161,$	$(1,20)$
$\Xi c(2815)2816.5\pm 1.2$	2808.5	$0,248668,$	$(1,20)$
$\Omega c0(2697)\pm 2.6$	2671.5	0.964034,	$(1,19)$
$\Lambda b0(5624)\pm 9$	5617	$(0.124467,)0$	$(2,20)$

Particle and mass	Mass From Formula	Error %	(l, n)
$\pi^{\pm}(139.57018) \pm 0.00035$	137	−1.43885	(2, 0)
$\pi^{0}(134.9766) \pm 0.0006$	137	1.481481	(2, 0)
$K^{\pm}4064 \pm 0.016$	496	1.9	(1, 3)
$\eta(547.51) \pm 0.18$	548	0.182815	(8, 0)
$f_0(600)(400 - 1200)$	616.5	(2.75)0	(1, 4)
$\rho(775.5) \pm 0.4$	753.5	−2.14286	(1.5)
$\omega(782) \pm 0.12$	753.5	−3.6445	(1, 5)
$\eta'(958) \pm 0.14$	959	0.104384	(2, 3)
$f_0(980) \pm 10$	959	−2.14286	(2, 3)
$a_0(980) \pm 1.2$	959	−2.14286	(2, 3)
$\phi(1020) \pm 0.020$	1027.5	0.735294	(1, 7)
$h_1(1170) \pm 20$	1164.5	(−0.47009)0	(1, 8)
$b_1(1235) \pm 3.2$	1233	(−0.16194)0	(2, 4)
$a_1(1260)1230 \pm 40$	1233	(−2.14286)0	(2, 4)
$f_2(1270)1275.4 \pm 1.1$	1233	−2.91339	(2, 4)
$f_1(1285)1281.8 \pm 0.6$	1301.5	1.284047	(1, 9)
$\eta(1295)1294 \pm 4$	1301.5	0.501931	(1, 9)
$\pi(1300) \pm 100$	1301.5	0.115385	(1, 9)
$a_2(1320)1318.3 \pm 0.6$	1301.5	−1.40152	(1, 9)
$f_0(1370)(1200 - 1500)$	1370	0	(4, 2)
$h_1(1380)$	1370	0.72464	(4, 2)
$\pi_1(1400)1376 \pm 17$	1370	−2.14286	(4, 2)
$f_1(1420)1426.3 \pm 0.9$	1438.5	1.302817	(1, 10)
$\omega(1420)1400 - 1450$	1438.5	(1.302817)0	(1, 10)
$f_2(1430)$	1438.5	0.594406	(1, 10)
$\eta(1440)(1400 - 1470)$	1438.5	−0.10417	(1, 10)
$a_0(1450)1474 \pm 19$	1438.5	−0.7931	(1, 10)
$\rho(1450)1459 \pm 11$	1438.5	−0.7931	(1, 10)
$f_0(1500)1507 \pm 5$	1507	(0.466667)0	(2, 5)
$f_1(1510)$	1507	−0.19868	(2, 5)
$f_2'(1525) \pm 5$	1507	−1.18033	(2, 5)
$f_2(1565)$	1575.5	0.670927	(1, 11)

Particle and mass	Mass From Formula	Error %	(l, n)
$h_1(1595)$	1575.5	-1.22257	$(1, 11)$
$\pi_1(1600)1653 \pm 8$	1575.5	-1.53125	$(1, 11)$
$\chi(1600)$	1575.5	-1.53125	$(1, 11)$
$a_1(1640)$	1644	0.243902	$(8, 1)$
$f_2(1640)$	1644	0.243902	$(8, 1)$
$\eta_2(1645)1617 \pm 5$	1644	(0.06079)0	$(8, 1)$
$\omega(1670)$	1644	(1.55688)0	$(8, 1)$
$\omega_3(1670)1667 \pm 4$	1644	-1.55689	$(8, 1)$
$\pi_2(1670)1672.4 \pm 3.2$	1644	-1.55689	$(8, 1)$
$\phi(1680) \pm 20$	1712.5	1.934524	$(1, 12)$
$\rho_3(1690)1688.8 \pm 2.1$	1712.5	1.331361	$(1, 12)$
$\rho(1700)1720 \pm 20$	1712.5	(0.735294)0	$(1, 12)$
$a_2(1700)$	1712.5	0.735294	$(1, 12)$
$f_0(1710)1718 \pm 6$	1712.5	(0.146199)0	$(1, 12)$
$\eta(1760)$	1781	1.193182	$(2, 6)$
$\pi(1800)1812 \pm 14$	1781	-1.05556	$(2, 6)$
$f_2(1810)$	1781	-1.60221	$(2, 6)$
$\phi_3(1850)1854 \pm 7$	1849.5	$(-0.02703)0$	$(1, 13)$
$\eta_2(1870)$	1849.5	-1.09626	$(1, 13)$
$\rho(1900)$	1918	0.947368	$(4, 3)$
$f_2(1910)$	1918	0.418848	$(4, 3)$
$f_2(1950)1944 \pm 12$	1918	-1.64103	$(4, 3)$
$\rho_3(1990)$	1986.5	-0.17588	$(1, 14)$
$X(2000)$	1986.5	-0.675	$(1, 14)$
$f_2(2010)(+60/-80)$	1986.5	$(-1.16915)0$	$(1, 14)$
$f_0(2020)$	1986.5	1.65842	$(1, 14)$
$a_4(2040)2001 \pm 10$	2055	0.735294	$(2, 7)$
$f_4(2050)2025 \pm 10$	2055	0.243902	$(2, 7)$
$\pi_2(2100)$	2123.5	1.119048	$(1, 15)$
$f_0(2100)$	2123.5	1.119048	$(1, 15)$
$f_2(2150)$	2123.5	-1.23256	$(1, 15)$

Particle and mass	Mass From Formula	Error %	(l, n)
$\rho_2(2150)$	2123.5	−1.23256	$(1, 15)$
$f_0(2200)$	2260.5	2.75	$(1, 16)$
$f_J(2220)$	2260.5	1.824324	$(1, 16)$
$\eta(2225)$	2360	1.595506	$(1, 16)$
$\rho_3(2250)$	2260	0.466667	$(1, 16)$
$f_2(2300)2297 \pm 28$	2329	1.26087	$(2, 8)$
$f_4(2300)$	2329	1.26087	$(2, 8)$
$D_s(2317)$	2329	0.5	$(2, 8)$
$f_0(2330)$	2329	−0.04292	$(2, 8)$
$f_2(2340)2339 \pm 60$	2329	−0.47009	$(2, 8)$
$\rho_5(2350)$	2329	−0.89362	$(2, 8)$
$a_6(2450)$	2466	−0.89362	$(4, 4)$
$f_6(2510)$	2534.5	0.976096	$(1, 18)$
$K^*(892) \pm 0.26$	890.5	−0.16816	$(1, 6)$
$K_1(1270)1272 \pm 7$	1233	2.91338	$(2, 4)$
$K_1(1400)1402 \pm 7$	1370	−2.14286	$(4, 2)$
$K^*(1410)1414 \pm 15$	1438.5	2.021277	$(1, 10)$
$K_0^*(1430)1414 \pm 6$	1438.5	0.594406	$(1, 10)$
$K_2^*(1430)1425.6 \pm 1.5$	1438.5	0.594406	$(1, 10)$
$K(1460)$	1438.5	−1.4726	$(1, 10)$
$Pentaquark(1.5GeV)$	1.5	0	$(2, 5)$
$K_2(1580)$	1575.5	−0.28481	$(1, 11)$
$K(1630)$	1644	0.858896	$(8, 1)$
$K_1(1650)$	1644	−0.36364	$(8, 1)$
$K^*(1680)1717 \pm 27$	1712.5	(1.934524)0	$(1, 12)$
$K_2(1770)1773 \pm 8$	1781	(0.621469)0	$(2, 6)$

Particle and mass	Mass From Formula	Error %	(l, n)
$K_3^*(1780) \pm 7$	1781	(0.05618)0	$(2, 6)$
$K_2(1820) \pm 13$	1849.5	1.620879	$(1, 13)$
$K(1830)$	1849.5	1.065574	$(1, 13)$
$K_0^*(1950)$	1918	-1.64103	$(4, 2)$
$K_2^*(1980)$	1986.5	0.328283	$(1, 14)$
$K_4^*(2045) \pm 9$	2055	(0.488998)0	$(2, 7)$
$K_2(2250)$	2260.5	0.466667	$(1, 16)$
$K_3(2320)$	2329	0.387931	$(2, 8)$
$K_5^*(2380)$	2397.5	0.735294	$(1, 17)$
$K_4(2500)$	2466	-1.36	$(4, 4)$
$K(3100)$	3082.5	-0.56452	$(1, 22)$
$D^{\pm}(1869.3)$	1849.5	-1.05922	$(1, 13)$
$D_0^{\pm}(1968.5) \pm 0.6$	1986.5	0.914402	$(1, 14)$
$D_0^*(2007)2006.7 \pm 0.4$	1986.5	-1.02143	$(1, 14)$
$D_{\pm}^*(2010) \pm 0.4$	1986.5	-1.16915	$(1, 14)$
$D_S(2317)2317.3 \pm 0.6$	2329	0.51791	$(2, 8)$
$D_1(2420)2422.3 \pm 1.3$	2397.5	-0.92975	$(1, 17)$
$D_1^{\pm}(2420)$	2397.5	-0.97067	$(1, 17)$
$D_2^*(2460)2461.1 \pm 1.6$	2466	0.243902	$(4, 4)$
$D_{\pm}^*(2460)2459 \pm 4$	2466	0.243902	$(4, 4)$
$D_{S1}^{\pm}(2536)2535.35 \pm 0.34$	2534.5	-0.07885	$(1, 18)$
$D_{SJ}(2573) \pm 1.5$	2534.5	-1.49631	$(1, 18)$
$B^{\pm}(5278)2579 \pm 0.5$	5274.5	-0.08524	$(1, 38)$
$B^0(5279.4) \pm 0.5$	5274.5	-0.09281	$(1, 38)$
$B_j(5732)$	5754	-0.47009	$(4, 10)$
$B_S^0(5369.6)5367.5 \pm 1.8$	5343	-0.49538	$(2, 19)$
$B_{SJ}^*(5850)$	5822.5	-0.47009	$(1, 42)$
$B_c^{\pm}(6400)6286 \pm 5$	6370.5	0.4609	$(3, 15)$
$\eta c(1S)(2979)2980.4 \pm 1.2$	2945.5	-1.12454	$(1, 21)$
$J/\psi(1S)(3096)3096.916 \pm 0.011$	3082.5	-0.46402	$(1, 22)$
$\chi c_0(1P)(3415.1)3414.76 \pm 0.35$	3425	0.289889	$(2, 12)$
$\chi c_1(1P)(3510.5) \pm 0.07$	3493.5	-0.48426	$(1, 25)$
$\chi c_2(1P)(3556)3556.20 \pm 0.09$	3562	0.168729	$(4, 6)$

Particle and mass	Mass From Formula	Error %	(l, n)
$\psi(2S)(3685.9)3686.093 \pm 0.034$	3699	0.355408	$(2, 13$
$\psi(3770)3771.1 \pm 2.4$	3767.5	$(-0.06631)0$	$(1, 27$
$\psi(3836)$	3836	0	$(8, 3$
$\chi(3872)3871.2 \pm 0.5$	3876	0.13	$(3, 9$
$\chi_{c2}(28)3929 \pm 5$	3944	0.38	$(2, 14$
$\psi(4040)4039 \pm 1$	4041.5	$(0.037129)0$	$(1, 29$
$\psi(4160)4153 \pm 3$	4178.5	$(0.444712)0$	$(1, 30$
$\psi(4415)4421 \pm 4$	4452.5	0.84937	$(1, 32$
$\gamma(1S)(9460.3) \pm 0.26$	9453	-0.07716	$(2, 34$
$\chi b_0(1P)(9859.9)9859.4 \pm 0.42 \pm 0.31$	9864	0.041583	$(16, 4$
$\chi b_1(1P)(9892.7) \pm 0.6$	9864	-0.29011	$(16, 4$
$\chi b_2(1P)(9912.6) \pm 0.5$	9864	-0.49029	$(16, 4$
$\gamma(2S)(10023) \pm 0.00031$	10001	0.21949	$(2, 3$
$\chi b_0(2P)(10232) \pm 0.0006$	10275	0.42026	$(2, 3$
$\chi b_1(2P)(10255) \pm 0.0005$	10275	0.1945027	$(2, 3$
$\chi b_2(2P)(10268) \pm 0.0004$	10275	0.068173	$(2, 3$
$\gamma(3S)(10355) \pm 0.0005$	10343.5	0.11105	$(1, 7$
$\gamma(4S)(10580)10579.4 \pm 1.2$	10549	-0.29301	$(2, 3$
$\gamma(10860)10865 \pm 8$	10891.5	0.290055	$(3, 2$
$\gamma(11020)11019 \pm 8$	11028.5	0.077132	$(1, 8$

Chapter 5

Spacetime Models and Tests

5.1 The Nature of Spacetime

Quantum Field Theory, General Relativity, Relativistic and Non Relativistic Quantum Theory and Classical Physics, all are based on a differentiable spacetime manifold, as indeed we have stressed earlier. This lead to an impasse, particularly in providing a unified description of Quantum phenomena and General Relativity. In particular, as Wheeler noted [45], there appeared no way of introducing the Quantum Mechanical spin half into General Relativity or the concept of curvature into Quantum Physics. In his words,
"the most evident shortcoming of the geometrodynamic model as it stands is this, that it fails to supply any completely natural place for spin half in general and for the neutrino in particular", while "it is impossible to accept any description of elementary particles that does not have a place for spin half."
So as we noted in Quantum Gravity including the author's approach, String Theory and other theories, be it at the Planck scale, or at the Compton scale we have turned to a non differentiable spacetime with minimum cut off intervals [315–317, 69]. The underlying reason for this breakdown of a differentiable spacetime manifold is the Uncertainty Principle—–as we go down to arbitrarily small spacetime intervals, we encounter arbitrarily large energy momenta. As Wheeler put it [45], "no prediction of spacetime, therefore no meaning for spacetime is the verdict of the Quantum Principle. That object which is central to all of Classical General Relativity, the four dimensional spacetime geometry, simply does not exist, except in a classical approximation."
Before proceeding to analyze the nature of spacetime beyond the classical

approximation, let us first analyze briefly the nature of classical spacetime itself. We can get an insight into the nature of the usual spacetime by revisiting the formulation of Quantum Theory in terms of stochastic processes more precisely, a Wiener process which, as we saw, models fuzzy spacetime [69, 318, 131, 130].
In the stochastic approach, we deal with a double Wiener process which leads to a complex velocity $V - \imath U$. As noted in Chapter 2 it is this complex velocity that leads to Quantum Theory from the usual diffusion theory.
To see this in a simple way, let us write the usual diffusion equation as

$$\Delta x \cdot \Delta x = \frac{h}{m} \Delta t \equiv \nu \Delta t \tag{5.1}$$

We saw that equation (5.1) can be rewritten as the usual Quantum Mechanical relation,

$$m \frac{\Delta x}{\Delta t} \cdot \Delta x = h = \Delta p \cdot \Delta x \tag{5.2}$$

We are dealing here, with phenomena within the Compton or De Broglie wavelength.
We now treat the diffusion constant ν to be very small, but non vanishing. That is, we consider the semi classical case. This is because, a purely classical description, does not provide any insight.
It is well known that in this situation we can use the WKB approximation [319]. Whence the right hand side of the wave function,

$$\psi = \sqrt{\rho} e^{\imath/\hbar S}$$

goes over to, in the one dimensional case, for simplicity,

$$(p_x)^{-\frac{1}{2}} e^{\frac{1}{\hbar}} \int p(x) dx$$

so that we have, on comparison,

$$\rho = \frac{1}{p_x} \tag{5.3}$$

ρ being the probability density. In this case the condition $U \approx 0$, that is, the velocity potential becoming real, implies

$$\nu \cdot \nabla ln(\sqrt{\rho}) \approx 0 \tag{5.4}$$

This semi classical analysis suggests that $\sqrt{\rho}$ is a slowly varying function of x, in fact each of the factors on the left side of (5.4) would be $\sim 0(h)$, so that the left side is $\sim 0(h^2)$ (which is being neglected). Then from (5.3) we conclude that p_x is independent of x, or is a slowly varying function of x. The equation of continuity now gives

$$\frac{\partial \rho}{\partial t} + \vec{\nabla}(\rho \vec{v}) = \frac{\partial \rho}{\partial t} = 0$$

That is the probability density ρ is independent or nearly so, not only of x but also of t. We are thus in a stationary and homogenous scenario. This is strictly speaking, possible only in a single particle Universe, or for a completely isolated particle, without any effect of the environment. Under these circumstances we have the various conservation laws and the time reversible theory, all this taken over into Quantum Mechanics as well. This is an approximation valid for small, incremental changes, as indeed is implicit in the concept of a differentiable spacetime manifold.
Infact the well known displacement operators of Quantum Theory which define the energy momentum operators are legitimate and further the energy and momenta are on the same footing only under this approximation[320]. If we retain terms $\sim O(l^2)$, where l is the minimum cut off, we encounter complex coordinates and spin half. Let us see this briefly.
In Dirac's theory of displacement operators [15] the operator $d_x \equiv \frac{d}{dx}$ is a purely imaginary operator, and is given by

$$\delta x(d_x + \bar{d}_x) = \delta x^2 d_x \bar{d}_x = 0$$

if

$$0(\delta x^2) = 0$$

as is tacitly assumed. However if

$$0(\delta x^2) \neq 0 \tag{5.5}$$

then the operator d_x becomes complex, and therefore, also the momentum operator, $p_x \equiv \imath\hbar d_x$ and the position operator. In other words if (5.5) holds good then we have to deal with complex or non-Hermitian coordinates. The implication of this is that (Cf.[299] for details) spacetime becomes non- commutative as seen earlier.
In any case here is the mysterious origin of the complex coordinates and spin [154]. The complex coordinates−−more specifically, an imaginary shift−−lead as we noted to the Kerr-Newman metric in the classical theory which describes the electron's field including the anomalous gyro magnetic ratio which are symptomatic of the electron's spin. It also means that the classical naked singularity in the Kerr-Newman metric is shielded by the fuzzy spacetime (Dirac's original averages over the zitterbewegung interval)

or equivalently the noncommutative geometry as we saw (Cf. also [60]).
We would now like to point out the well known close similarity between the stochastic formulation mentioned above and discussed in Chapter 2 and the hydrodynamical formulation for Quantum Mechanics, which also leads to identical equations on writing the wave function as above. These two approaches were reconciled by considering quantized vortices at the Compton scale (Cf.[130, 127]).
To proceed further, we start with the Schrödinger equation

$$\imath\hbar\frac{\partial\psi}{\partial t} = -\frac{\hbar^2}{2m}\nabla^2\psi + V\psi \tag{5.6}$$

Remembering that for momentum eigen states we have, for simplicity, for one dimension

$$\frac{\hbar}{\imath}\frac{\partial}{\partial x}\psi = p\psi \tag{5.7}$$

where p is the momentum or p/m is the velocity v, we take the derivative with respect to x (or $\vec{x}$) of both sides of (5.6) to obtain, on using (5.7),

$$\imath\hbar\frac{\partial(v\psi)}{\partial t} = -\frac{\hbar^2}{2m}\nabla^2(v\psi) + \frac{\partial V}{\partial x}\psi + Vv\psi \tag{5.8}$$

We would like to compare (5.8) with the well known equation for the velocity in hydrodynamics[321, 322], following from the Navier-Stokes equation,

$$\rho\frac{\partial v}{\partial t} = -\nabla p - \rho\alpha T g + \mu\nabla^2 v \tag{5.9}$$

In (5.9) v is a small perturbational velocity in otherwise stationary flow between parallel plates separated by a distance d, p is a small pressure, ρ is the density of the fluid, T is the temperature proportional to $Q(z)v, \mu$ is the Navier-Stokes coefficient and α is the coefficient of volume expansion with temperature. Also required would be

$$\beta \equiv \frac{\Delta T}{d}.$$

v itself is given by

$$v_z = W(z)exp(\sigma t + \imath k_x x + \imath k_y y), \tag{5.10}$$

z being the coordinate perpendicular to the fluid flow.
We can now see the parallel between equations (5.8) and (5.9). To verify the identification we would require that the dimensionless Rayleigh Number

$$R = \frac{\alpha\beta g d^4}{\kappa\nu}$$

should have an analogue in (5.8) which is dimensionless, κ, ν being the thermometric conductivity and viscosity.
Remembering that

$$\nu \sim \frac{h}{m}$$

and

$$d \sim \lambda$$

where λ is the Compton wavelength in the above theory (Cf.[130, 129] for details) and further we have

$$\rho \propto f(z)g = V \qquad (5.11)$$

for the identification between the hydrostatic energy and the energy V of Quantum Mechanics, it is easy using (5.11) and earlier relations to show that the analogue of R is

$$(c^2/\lambda^2) \cdot \lambda^4 \cdot (m/h)^2 \qquad (5.12)$$

The expression (5.12) indeed is dimensionless and of order 1. Thus the mathematical identification is complete.
Before proceeding, let us look at the physical significance of the above considerations (Cf.[5] for a graphic description.) Under conditions of stationary flow, when the difference in the temperature between the two plates is negligible there is total translational symmetry, as in the case of the displacement operators of Quantum Theory. But when there is a small perturbation in the velocity (or equivalently the temperature difference), then beyond a critical value the stationarity and homogeneity of the fluid is disrupted, or the symmetry is broken and we have the phenomenon of the formation of Benard cells, which are convective vortices and can be counted. This infact is the "birth" of space It must be stressed that before the formation of the Benard cells, there is no "space", that is, no point to distinguish from or relate to another point. Only with the formation of the cells are we able to label space points.
In the context of the above identification, the Benard cells would correspond to the formation of "quantized vortices" at the Compton (Planck) scale from the ZPF, as we saw, which latter had been discussed in detail in the literature (Cf.[130] and [323]). This phase transition discussed from the Landau-Ginzburg theory earlier, would correspond to the "formation" of spacetime. As discussed in detail in [130] these "quantized vortices" can be identified with elementary particles. Interestingly, as noted Einstein himself considered electrons as condensates from a background electromagnetic field.

However in order to demonstrate that the above formulation is not a mere mathematical analogy, we have to show that the critical value of the wave number k in the expression for the velocity in the hydrodynamical flow (5.10) is the same as the critical length, the Compton length. In terms of the dimensionless wave number $k' = k/d$, this critical value is given by[321]

$$k'_c \sim 1$$

In the case of the "quantized vortices" at the Compton scale l, remembering that d is identified with l itself we have,

$$l'_c (\equiv) k'_c \sim 1,$$

exactly as required.
In this connection it may be reiterated that due to fluctuations in the Zero Point Field or the Quantum vacuum, there would be fluctuations in the metric given by[45]

$$\Delta g \sim l_P/l$$

where l_P is the Planck length and l is a small interval under consideration. At the same time the fluctuation in the curvature of space would be given by

$$\Delta R \sim l_P/l^3$$

Normally these fluctuations are extremely small but as discussed in detail elsewhere[175], this would imply that at the Compton scale of a typical elementary particle $l \sim 10^{-11} cms$, the fluctuation in the curvature would be ~ 1. This is symptomatic of the formation of what we have termed above as elementary particle "quantized vortices".
Further if a typical time interval between the formation of such "quantized vortices" which are the analogues of the Benard cells is τ, in this case the Compton time, then as in the theory of the Brownian Random Walk[108], the mean time extent would be given by, as noted earlier,

$$T \sim \sqrt{N}\tau \tag{5.13}$$

where N is the number of such quantized vortices or elementary particles (Cf.also [130, 127]). The equation (5.13) holds good for the Universe itself because T the age of the Universe $\sim 10^{17} secs$ and N the number of elementary particles $\sim 10^{80}$, τ being the Compton time $\sim 10^{-23} secs$. Interestingly, this "phase transition" nature of time would automatically make it irreversible, unlike the conventional model in which time is reversible.

We will return to these considerations later in this section.
As we saw an equation similar to (5.13) can be deduced by the same arguments for space extension also as indeed we did earlier, and this time we get back the well known Eddington formula viz.,

$$R \sim \sqrt{N} l \tag{5.14}$$

where R is the extent or radius of the Universe and l is the cell size or Compton wavelength. We can similarly characterize the formation of elementary particles themselves from cells at the Planck scale.
Once we recognize the minimum spacetime extensions, then we immediately are lead to the underlying non commutative geometry encountered in earlier chapters and given by

$$[x, y] = 0(l^2), [x, p_x] = \imath\hbar[1 + 0(l^2)], [t, E] = \imath\hbar[1 + 0(\tau^2)] \tag{5.15}$$

As we noted relations like (5.15) are Lorentz invariant. At this stage we recognise the new nature of spacetime as given by (5.15) in contrast to the stationary and homogeneous spacetime discussed earlier. Witten [243, 166] has called this Fermionic spacetime as contrasted with the usual Bosonic spacetime. Indeed we will trace later on, the origins of the Dirac equation of the electron to (5.15). We will also argue that (5.15) leads to the long sought after reconciliation between Electromagnetism and Gravitation in an extended gauge formulation [299, 324].
The usual differentiable spacetime geometry can be obtained from (5.15) if l^2 is neglected, and this is the approximation that has been implicit. Retaining terms of the order of l leads us to Quantum Theory including Quantum Field Theory. Finally if we neglect $O(l)$, too, then we are lead to Classical Physics (including General Relativity).
Thus spacetime is a collection of such cells or elementary particles. As pointed out earlier, this spacetime emerges from a homogeneous stationary non or pre spacetime when the symmetry is broken, through random processes. The question that comes up then is, what is the metric which we use? We will examine it now.
We first make a few preliminary remarks. When we talk of a metric or the distance between two "points" or "particles", a concept that is implicit is that of topological "nearness"—–we require an underpinning of a suitably large number of "open" sets [325]. Let us now abandon the absolute or background spacetime and consider, for simplicity, a Universe (or set) that consists solely of two particles. The question of the distance between these particles (quite apart from the question of the observer) becomes meaningless. Indeed, this is so for a Universe consisting of a finite number of

particles. For, we could isolate any two of them, and the distance between them would have no meaning. We can intuitively appreciate that we would in fact need distances of intermediate or more generally, other points.
In earlier work[326, 197], motivated by physical considerations we had considered a series of nested sets or neighborhoods which were countable and also whose union was a complete Hausdorff space. The Urysohn Theorem was then invoked and it was shown that the space of the subsets was metrizable. Let us examine this in more detail.
Firstly we observe that in the light of the above remarks, the concepts of open sets, connectedness and the like reenter in which case such an isolation of two points would not be possible. More formally let us define a neighborhood of a particle (or point or element) A of a set of particles as a subset which contains A and atleast one other distinct element. Now, given two particles (or points) or sets of points A and B, let us consider a neighborhood containing both of them, $n(A, B)$ say. We require a non empty set containing at least one of A and B and at least one other particle C, such that $n(A, C) \subset n(A, B)$, and so on. Strictly, this "nested" sequence should not terminate. For, if it does, then we end up with a set $n(A, P)$ consisting of two isolated "particles" or points, and the "distance" $d(A, P)$ is meaningless.
We now assume the following property[326, 197]: Given two distinct elements (or even subsets) A and B, there is a neighborhood N_{A_1} such that A belongs to N_{A_1}, B does not belong to N_{A_1} and also given any N_{A_1}, there exists a neighborhood $N_{A_{\frac{1}{2}}}$ such that $A \subset N_{A_{\frac{1}{2}}} \subset N_{A_1}$, that is there exists an infinite topological closeness.
From here, as in the derivation of Urysohn's Lemma[327], we could define a mapping f such that $f(A) = 0$ and $f(B) = 1$ and which takes on all intermediate values. We could now define a metric, $d(A, B) = |f(A) - f(B)|$. We could easily verify that this satisfies the properties of a metric.
With the same motivation we will now deduce a similar result, but with different conditions. In the sequel, by a subset we will mean a proper subset, which is also non null, unless specifically mentioned to be so. We will also consider Borel sets, that is the set itself (and its subsets) has a countable covering with subsets. We then follow a pattern similar to that of a Cantor ternary set [325, 328]. So starting with the set N we consider a subset N_1 which is one of the members of the covering of N and iterate this process so that N_{12} denotes a subset belonging to the covering of N_1 and so on.
We note that each element of N would be contained in one of the series

of subsets of a sub cover. For, if we consider the case where the element p belongs to some $N_{12\cdots m}$ but not to any $N_{1,2,3\cdots m+1}$, this would be impossible because the latter form a cover of the former. In any case as in the derivation of the Cantor set, we can put the above countable series of sub sets of sub covers in a one to one correspondence with suitable sub intervals of a real interval (a, b).

Case I

If $N_{1,2,3\cdots m} \to$ an element of the set N as $m \to \infty$, that is if the set is closed, we would be establishing a one to one relationship with points on the interval (a, b) and hence could use the metric of this latter interval, as seen earlier.

Case II

It is interesting to consider the case where in the above iterative countable process, the limit does not tend to an element of the set N, that is set N is not closed and has what we may call singular points. We could still truncate the process at $N_{1,2,3\cdots m}$ for some $m > L$ arbitrary and establish a one to one relationship between such truncated subsets and arbitrarily small intervals in a, b. We could still speak of a metric or distance between two such arbitrarily small intervals.

This case which may be termed "Fuzzy Topology", is of interest because of our description of elementary particles in terms of fuzzy spacetime (Cf. also ref.[130]), where we have a length of the order of the Compton wavelength as seen in the previous sections, within which spacetime as we know it breaks down. Such cut offs as seen lead to a non commutative geometry (5.15) and what may be called fuzzy spaces[329, 45].(We note that the centre of the vortex in the above model is a singular point). In any case, the number of particles in the Universe is of the order 10^{80}, which approximates infinity from a physicist's point of view.

Interestingly, we usually consider two types of infinite sets––those with cardinal number n corresponding to countable infinities, and those with cardinal number c corresponding to a continuum, there being nothing in between [328]. This is the well known but unproven Continuum hypothesis. What we have shown with the above process is that it is possible to conceive of an intermediate possibility with a cardinal number $n^p, p > 1$.

In the above considerations three properties are important: Firstly the set must be closed i.e. it must contain all its limit points. Secondly, it must be perfect i.e. in addition, each of its points must be a limit point. Finally it must be disconnected i.e. it contains no non null open intervals. Only the first was invoked in Case I.

We notice that there is the holistic feature. A metric emerges by considering large encompassing sets. Finally, we could deviate from a strict mathematical analysis and introduce an element of physics. We could say that a point or particle B would be in a neighborhood of another point or particle A, only if A and B interact". Thus the universe would consist of a network of "interacting" particles, reminiscent of the Feynman-Wheeler perfect absorber model encountered in Chapter 1.

It may be remarked that much of Quantum Theory, like much of Classical Theory was couched in the concepts of Newtonian two body mechanics and determinism. The moment we consider even a three body problem, as was realized by Poincare more than a century ago and noted in Chapter 1, the picture gets altered. As he noted [330],

"If we knew exactly the laws of nature and the situation of the Universe at the initial moment, we could predict exactly the situation of that same Universe at a succeeding moment. But even if it were the case that the natural laws had no longer any secret for us, we could still know the situation approximately. If that enabled us to predict the succeeding situation with the same approximation, that is all we require, and we should say that the phenomenon had been predicted, that it is governed by the laws. But it is not always so; it may happen that small differences in the initial conditions produce very great ones in the final phenomena. A small error in the former will produce an enormous error in the latter. Prediction becomes impossible."

In a similar vein, Prigogine observes as we noted, [5], "Our physical world is no longer symbolized by the stable and periodic planetary motions that are at the heart of classical mechanics. It is a world of instabilities and fluctuations..."

Indeed, the departure from the two body formulation began with Electromagnetism itself, which has to invoke the environment. However, a nuance must be brought up. This realization of chaos in the physical universe, is still compatible, and in fact has been worked out in, the usual Newtonian space. We on the other hand, are emphasizing the chaotic or stochastic feature of spacetime itself.

We now return to the current view of Planck scale oscillators in the background dark energy or Quantum vacuum. In this context we saw in the last Chapter that elementary particles can be considered to be normal modes of $n \sim 10^{40}$ Planck oscillators in the ground state, while the entire Universe itself has an underpinning of $\bar{N} \sim 10^{120}$ such oscillators, there being $N \sim 10^{80}$ elementary particles in the Universe [157, 158]. These Planck oscillators are formed out of the Quantum vacuum (or dark energy, as we repeatedly emphasized). Thus we noticed that, $m_P c^2$ being the energy of each Planck oscillator, m_P being the Planck mass, $\sim 10^{-5} gms$,

$$m = \frac{m_P}{\sqrt{n}} \tag{5.16}$$

$$l = \sqrt{n} l_P, \tau = \sqrt{n} \tau_P, n = \sqrt{N} \tag{5.17}$$

where m is the mass of a typical elementary particle, taken to be a pion in the literature. The ground state of $\bar{N}$ such Planck oscillators would be, in analogy to (5.16),

$$m_\gamma = \frac{m_P}{\sqrt{\bar{N}}} \sim 10^{-65} gms \tag{5.18}$$

This we had identified with the mass of the photon. The Universe is an excited state and consists of N such ground state levels or elementary particles and so we have, from (5.18)

$$M = mN = \sqrt{\bar{N}} m_P \sim 10^{55} gms,$$

as required, M being the mass of the Universe. It must be noted that M is expressible in two different ways−−one in terms of the elementary particle mass and the other in terms of the Planck mass. In the latter case, we have a coherent collection of $\bar{N}$ Planck oscillators. In the former, we have a collection of N, what may be called, disentangled elementary particles. Hence the difference. We will come back to this nuance soon.
Due to the fluctuation $\sim \sqrt{n}$ in the levels of the n oscillators making up an elementary particle, the resulting energy is, remembering that mc^2 is the ground state,

$$\Delta E \sim \sqrt{n} mc^2 = m_P c^2,$$

using (5.17), and so the indeterminacy time is

$$\frac{\hbar}{\Delta E} = \frac{\hbar}{m_P c^2} = \tau_P,$$

as indeed we would expect.
The corresponding minimum indeterminacy length would therefore be l_P. One of the consequences of the minimum spacetime cut off is that the Heisenberg Uncertainty Principle takes an extra term as we saw in the previous Chapter [331]. Thus as we saw

$$\Delta x \approx \frac{\hbar}{\Delta p} + \alpha \frac{\Delta p}{\hbar}, \ \alpha = l^2 (\text{or}\, l_P^2) \tag{5.19}$$

where l (or l_P) is the minimum interval under consideration. The first term gives the usual Heisenberg Uncertainty Principle.
Application of the time analogue of (5.19) for the indeterminacy time Δt for the fluctuation in energy $\Delta \bar{E} = \sqrt{N} mc^2$ in the N particle states of the Universe gives exactly as above,

$$\Delta t = \frac{\Delta E}{\hbar} \tau_P^2 = \frac{\sqrt{N} mc^2}{\hbar} \tau_P^2 = \frac{\sqrt{N} m_P c^2}{\sqrt{n} \hbar} \tau_P^2 = \sqrt{n} \tau_P = \tau,$$

wherein we have used (5.17). In other words, for the fluctuation $\sqrt{N}$, the time is τ as indeed we saw in Chapters 2 and 3. It must be re-emphasized that the Compton time τ of an elementary particle, is an interval within which there are unphysical effects like zitterbewegung-–as pointed out by Dirac, an averaging over this interval is required. This gives us,

$$dN/dt = \sqrt{N}/\tau \tag{5.20}$$

Equation (5.20) was the starting point for the cosmology discussed in Chapter 3. Here we have re-derived this relation from a consideration of the underlying Planck oscillators. On the other hand due to the fluctuation in the $\sqrt{\bar{N}}$ oscillators constituting the Universe, the fluctuational energy is similarly given by

$$\sqrt{\bar{N}} m_P c^2,$$

which gives the mass of the universe. As we already saw, this shows that like spacetime, the matter of the universe too is a dispersion or fluctuation in the background dark energy or ZPF.
Another way of deriving (5.20) is to observe that as $\sqrt{n}$ oscillators appear fluctuationally in time τ_P which translates, in the elementary particle time scales, $\sqrt{n}\sqrt{n} = \sqrt{N}$ particles in $\sqrt{n}\tau_P = \tau$, the rate of the fluctuational appearance of particles is

$$\left(\frac{\sqrt{n}}{\tau_P} \right) = \frac{\sqrt{N}}{\tau} = dN/dt$$

which is again (5.20). From here by integration,

$$T = \sqrt{N} \tau$$

T is the time elapsed from $N = 1$ and τ is the Compton time. This gives T its origin in the fluctuations-–there is no smooth "background" (or "being") time-–as noted, the root of time is in "becoming". It is the time of a Brownian or double Wiener process: A step l gives a step in time $l/c \equiv \tau$ and therefore $\Delta x = \sqrt{N} l$ gives $T = \sqrt{N} \tau$. Time is born out of acausal

fluctuations which are random and therefore irreversible. Indeed, there is no background time. Time is proportional to $\sqrt{N}$, (N being the number of particles) which are being created spontaneously from the ZPF. It is in this sense that spacetime and matter are a dispersion. We already had argued for this in Chapter 2. To emphasize further, if there were N steps of length l, the usual displacement would be Nl and not $\sqrt{N}l$.
On the other hand the time we actually use is what may be called stationary time and it is an approximation as we saw earlier [325].
Further, Quantum Mechanics, Gravitation etc. follow from here. In Quantum Mechanics, the measurement of the observer triggers the acausal collapse of the wave function−−an irreversible event−−but the wave function itself as noted in Chapter 2, satisfies a deterministic and reversible equation, paradoxically. Yet the Universe is "irreversible". It appears spontaneous irreversibility or indeterministic time [127] is the real time. This can be contrasted to the usual time reversibility embedded in Newtonian dynamics and thence Quantum Theory. So, if to put it simply, $t \to -t$ (or we use the Time Reversal Operator), the laws of physics remain invariant. Specifically, in the case of Quantum Theory, we observe that the wave function is given by [144]

$$\psi(r,t) = \frac{1}{(2\pi\hbar)^{3/2}} \int a(p) \exp\left[\frac{\imath}{\hbar}\left(p\cdot r - \frac{p^2}{2m}t\right)\right] dp,$$

$a(p)$, the amplitude in momentum space being independent of time. So we have at any other time t':

$$a(p) = \frac{1}{(2\pi\hbar)^{3/2}} \int \psi(r',t') \exp\left[-\frac{\imath}{\hbar}\left(p\cdot r' - \frac{p^2}{2m}t'\right)\right] dr'$$

Substitution of $a(p)$ yields the result

$$\psi(r,t) = \int K(r,t';r',t')\psi(r',t')dr', \tag{5.21}$$

the Kernel function K being given by

$$K(r,t;r',t') = \frac{1}{(2\pi\hbar)^3} \int \exp\left\{\frac{\imath}{\hbar}\left[p\cdot(r-r') - \frac{p^2}{2m}(t-t')\right]\right\} dp$$

or after some manipulation, in the form

$$K(r,t;r',t') = \left[\frac{2\pi\imath\hbar}{m}(t-t')\right]^{-3/2} \exp\left[\imath\frac{m}{2\hbar}\frac{|r-r'|^2}{(t-t')}\right]$$

The point is that in (5.21) $\psi(r,t)$ at t is given in terms of a linear expansion of $\psi(r,t')$ at earlier times t'. But what is to be noted is, the symmetry between t and t'. This is not surprising as the original Schrödinger equation itself remains unchanged under $t \to -t$. On the contrary, as we saw in Chapter 2, in a diffusion process, this symmetry between past and future ceases.

Thus it is possible to understand the fluctuations, that is, the equation (5.20) which was the starting point for fluctuational energy in terms of the underpinning of Planck scale oscillators in the Quantum vacuum or dark energy.

We would now like to reiterate the following. Starting from a completely different point of view namely Black Hole Thermodynamics, Landsberg [254] as noted deduced that the smallest observable mass in the Universe is $\sim 10^{-65} gms$, which is exactly the minimum mass encountered earlier and given in (5.18).

Now due to the fluctuational appearance of $\sqrt{N}$ particles, the fluctuational mass associated with each of the N particles in the Universe is

$$\frac{\sqrt{N}m}{N} = \frac{m}{\sqrt{N}} \sim 10^{-65} gms,$$

that is once again the smallest observable mass or ground state mass in the Universe.

We have already argued that there must be fluctuations of the Quantum Electromagnetic Field, as required by the Heisenberg Principle, so that we have for an extent $\sim L$ (B being the magnetic field), the energy density

$$(\Delta B)^2 \geq \hbar c / L^4 \tag{5.22}$$

Whence from (5.22), the energy in the entire volume $\sim L^3$ is given by, as we saw in Chapter 4,

$$\Delta E \sim \hbar c / L \tag{5.23}$$

From another angle Braffort and coworkers deduced the Zero Point Field from the Absorber Theory of Wheeler and Feynman, which we encountered earlier in Chapter 1. In the process they found that the spectral density of the vacuum field was given by [80, 332]

$$\rho(\omega) = \text{const} \cdot \omega^3 \tag{5.24}$$

This result also follows from (5.22) or (5.23), remembering that $\omega = c/L$, as we saw earlier in Chapter 3. There have been other points of view which converge to the above conclusions. In any case as we have seen a little earlier, it would be too much of an idealization to consider an atom or a charged particle to be an isolated system. It is the interaction with the rest

of the universe that produces a random field.
It has also been shown that the constant of proportionality in (5.24) is given by (Cf.ref.[80])

$$\frac{\hbar}{2\pi^2 c^3}$$

a result which can be obtained also from (5.22) or (5.23). Interestingly such a constant is implied by Lorentz invariance.
From the point of view of Quantum Electrodynamics we reach conclusions similar to those seen above. As Feynman and Hibbs put it [122]
"Since most of the space is a vacuum, any effect of the vacuum-state energy of the electromagnetic field would be large. We can estimate its magnitude. First, it should be pointed out that some other infinities occurring in quantum-electrodynamic problems are avoided by a particular assumption called the *cut off rule.* This rule states that those modes having very high frequencies (short wavelength) are to be excluded from consideration. The rule is justified on the ground that we have no evidence that the laws of electrodynamics are obeyed for wavelengths shorter than any which have yet been observed. In fact, there is a good reason to believe that the laws cannot be extended to the short-wavelength region.
"Mathematical representations which work quite well at longer wavelengths lead to divergences if extended into the short wavelength region. The wavelengths in question are of the order of the Compton wavelength of the proton; $1/2\pi$ times this wavelength is $\hbar/mc \simeq 2 \times 10^{-14} cm$.
"For our present estimate suppose we carry out sums over wave numbers only up to the limiting value $k_{max} = mc/\hbar$. Approximating the sum over states by an integral, we have, for the vacuum-state energy per unit volume,

$$\frac{E_e}{\text{unit} \quad \text{vol}} = 2\frac{\hbar c}{2(2\pi)^3} \int_0^{k_{max}} k 4\pi k^2 dk = \frac{\hbar c k_{max}^4}{8\pi^2}$$

"(Note the first factor 2, for there are two modes for each k). The equivalent mass of this energy is obtained by dividing the result by c^2. This gives

$$\frac{m_0}{\text{unit} \quad \text{vol}} = 2 \times 10^{15} g/cm^3$$

"Such a mass density would, at first sight at least, be expected to produce very large gravitational effects which are not observed. It is possible that we are calculating in a naive manner, and, if all of the consequences of the general theory of relativity (such as the gravitational effects produced by the large stresses implied here) were included, the effects might cancel out; but nobody has worked all this out. It is possible that some cutoff

procedure that not only yields a finite energy density for the vacuum state but also provides relativistic invariance may be found. The implications of such a result are at present completely unknown.
"For the present we are safe in assigning the value zero for the vacuum-state energy density. Up to the present time no experiments that would contradict this assumption have been performed."
However the high density encountered above is perfectly meaningful if we consider the Compton scale cut off $\sim 10^{-13}cm$: Within this volume the density gives us back the mass of an elementary particle like the pion as already seen.

5.2 Other Formulations

We now consider alternative formulations which retain the spirit of a "thermodynamic" spacetime. We start with our view of spacetime as a collection of Planck oscillators.
Let us denote the state of each Planck oscillator by ϕ_n; then the state of the Universe can be described in the spirit of entanglement discussed earlier by

$$\psi = \sum_n c_n \phi_n, \tag{5.25}$$

So ψ represents the state of the universe in terms of coherent Planck oscillators and ϕ_n which can be considered to be eigen states of energy with eigen values E_n. It is known that (5.25) can be written as [333]

$$\psi = \sum_n b_n \bar{\phi}_n \tag{5.26}$$

where $|b_n|^2 = 1$ if $E < E_n < E + \Delta$ and $= 0$ otherwise under the assumption

$$\overline{(c_n, c_m)} = 0, n \neq m \tag{5.27}$$

(In fact n in (5.27) could stand for not a single state but for a set of states $n_\imath$, and so also m.) Here the bar denotes a time average over a suitable interval. This is the well known Random Phase Axiom and arises due to the total randomness amongst the phases c_n. We have already encountered, essentially equation (5.27), in the previous Chapter––these random jigglings in the ZPF, as we saw, averaged off over the real life Compton scale.
We stress that the difference between (5.25) and (5.26) is that in the latter, the "eigen" states (or energy) are not so sharply defined. This is the real

world scenario−−rather than the Planck scale, described by (5.25). The expectation value of any operator O is given by

$$< O >= \sum_n |b_n|^2(\bar{\phi}_n, O\bar{\phi}_n)/\sum_n |b_n|^2 \qquad (5.28)$$

Equation (5.28) is as in the usual theory of Quantum Mechanical operators. Equations (5.27) and (5.26) show that effectively we have incoherent states $\bar{\phi}_1, \bar{\phi}_2, \cdots$ once averages over time intervals for the phases c_n in (5.27) vanish owing to their relative randomness. This is in contrast to the entangled states in (5.25). It was in this sense that, as we saw a little earlier the mass of the universe is given in terms of $\sqrt{N}$ Planck oscillators, but N elementary particles. All this has to be viewed in the perspective of the discussion of Schrondinger's Cat in Chapter 1.
In the light of the preceding discussion of random fluctuations, we can interpret all this meaningfully: We can identify ϕ_n with the ZPF. The time averages are the same as Dirac's zitterbewegung averages over intervals $\sim \frac{\hbar}{mc^2}$ this being due to the presence of $\sim 10^{40}$ Planck oscillators per elementary particle, bunched together (Cf.ref.[130]). We then get disconnected or incoherent particles from a single background of vacuum fluctuations exactly as before. The incoherence arises because of the well known random phase relation (5.27), that is after averaging over the suitable interval, this again being symptomatic of the unphysical (zitterbewegung) processes within the Compton scale. Here the entanglement is weakened by the interactions and hence we have (5.26) for elementary particles, rather than (5.25) (Cf.ref.[274]).
Let us look at this from another point of view. We observe that the coherent N' Planck oscillators referred to above could be considered to be a degenerate Bose assembly. In this case as is well known we have

$$v = \frac{V}{N'}$$

(Cf.ref.[125]−−here z of the usual theory ≈ 1). V the volume of the universe $\sim 10^{84} cm^3$. Whence

$$v = \frac{V}{N'} \sim 10^{-36}$$

So that the wavelength

$$\lambda \sim (v)^{1/3} \sim 10^{-12} cm = l \qquad (5.29)$$

What is very interesting is that (5.29) gives us the Compton length of a typical elementary particle like the pion. So from the Planck oscillators we are able to recover the elementary particles exactly as before [157, 158]. So our description of the universe at the Planck scale is that of an entangled wave function as in (5.25). However we perceive the universe at the elementary particle or Compton scale, where the random phases would have weakened the entanglement, and we have the description as in (5.26). Does this mean that N elementary particles in the universe are totally incoherent in which case we do not have any justification for treating them to be in the same spacetime?

We can argue that they still interact amongst each other though in comparison this is "weak". For instance let us consider the background ZPF whose spectral frequency is given by (5.24). If there are two particles at A and B separated by a distance r, then those wavelengths of the ZPF which are at least $\sim r$ would connect or link the two particles. Whence the force of interaction between the two particles is given by, remembering that $\omega \propto 1/r$,

$$\text{Force} \propto \int_r^\infty \omega^3 dr \propto \frac{1}{r^2} \tag{5.30}$$

Thus from (5.30) we are able to recover the familiar Coulomb Law of interaction. The background ZPF thus enables us to recover the action at a distance formulation. We had discussed this in Chapter 1. In fact a similar argument can be given [334] to recover from QED the Coulomb Law––here the carriers of the force are the virtual photons, that is photons whose life time is within the Compton time of uncertainty permitted by the Heisenberg Uncertainty Principle.

It is thus possible to synthesize the field and action at a distance concepts, once it is recognized that there is the ZPF and there are minimum spaetime intervals at the Compton scale [14]. Many of the supposed contradictions arise because of our characterization in terms of spacetime points and a differentiable manifold. Once the minimum cut off at the Planck scale is introduced, this leads to the physical Compton scale and a unified formulation free of divergence problems. We now make a few comments.

We had seen in Chapter 1 that the Dirac formulation of Classical Electrodynamics needed to introduce the acausal advanced field. However the acausality was again within the Compton time scale. In fact this fuzzy spacetime was modelled by a Wiener process as discussed in Chapters 2 and 4 [16] (Cf. also ref.[131]). The point here to recapitulate, is that the backward and forward time derivatives for $\Delta t \to 0^-$ and 0^+ respectively do not cancel, as they should not, if time is fuzzy. So we automatically recover from the electromagnetic potential the retarded field for forward derivatives and the advanced fields for backward derivatives. In this case we have to consider both these fields. Causality however is recovered as we saw. This is a transition to intervals which are greater in magnitude compared to the Compton scale.

It must also be mentioned that a few assumptions are implicit in the conventional theory using differentiable spacetime manifolds. In the variational problem we use the conventional δ (variation) which commutes with the time derivatives. So such an operator is constant in time. So also the energy momentum operators in Dirac's displacement operators theory are the usual time and space derivatives of Quantum Theory. But here the displacements are "instantaneous". They are valid in a stationary or constant energy scenario, and it is only then that the space and time operators are on the same footing as required by Special Relativity [320]. In fact we have argued that in this theory we neglect intervals $\sim 0(\delta x^2)$. But if δx is of the order of the Compton scale and we do not neglect the square of this scale, then the space and momentum coordinates become complex indicative of a noncommutative geometry which has been discussed in detail earlier [154, 335, 16].

What all this means is that it is only on neglecting $0(l^2)$ that we have the conventional spacetime of Quantum Theory, including relativistic Quantum Mechanics and Special Relativity, that is the Minkowski spacetime. Coming to the conservation laws of energy and momentum these are based on translation symmetries [49]−−what it means is the operators $\frac{d}{dx}$ or $\frac{d}{dt}$ are independent of x and t. There is here a homogeneity property of spacetime which makes these laws in effect non local. This has to be borne in mind, particularly when we try to explain the EPR paradox encountered in Chapter 1.

We noted that John Wheeler had stressed that the divide between Classical and Quantum Theory lies in the spin half (of Fermions) [45]. This half integral spin gives rise to such non classical and purely Quantum Mechanical results as the anomalous gyromagnetic ratio of the electron ($g = 2$). We will now argue that the non classical half integral spin feature arises from the multiply connected nature of Quantum Spacetime, and it is this which distinguishes Quantum Mechanics from Classical Theory. Specifically, we will argue that the usual space R with a compactified space S^1, in $R \times S^1$ reproduces Quantum Mechanical spin. On the other hand, spacetime is simply connected in Classical theory.

5.3 Multiply Connected Space and Spin

Let us start by reviewing Dirac's original derivation of the Monopole (Cf.ref.[51]). He started with the wave function

$$\psi = Ae^{\imath\gamma}, \tag{5.31}$$

He then considered the case where the phase γ in (5.31) is non integrable. In this case (5.31) can be rewritten as

$$\psi = \psi_1 e^{\imath S}, \tag{5.32}$$

where ψ_1 is an ordinary wave function with integrable phase, and further, while the phase S does not have a definite value at each point, its four gradient viz.,

$$K^\mu = \partial^\mu S \tag{5.33}$$

is well defined. We use temporarily natural units, $\hbar = c = 1$. Dirac then goes on to identify K in (5.33) (except for the numerical factor hc/e) with the electromagnetic field potential, as in the Weyl gauge invariant theory. Next Dirac considered the case of a nodal singularity, which is closely related to what was later called a quantized vortex a term we have already noted (Cf. for example ref.[323]). In this case a circuit integral of a vector as in (5.33) gives, in addition to the electromagnetic term, a term like $2\pi n$, so that we have for a change in phase for a small closed curve around this nodal singularity,

$$2\pi n + e\int \vec{B}\cdot d\vec{S} \tag{5.34}$$

In (5.34) $\vec{B}$ is the magnetic flux across a surface element $d\vec{S}$ and n is the number of nodes within the circuit. The expression (5.34) directly lead to the Monopole in Dirac's formulation.

Let us now reconsider the above arguments in terms of our earlier developments. As we saw the Dirac equation for a spin half particle throws up a complex or non Hermitian position coordinate [158, 154]. Dirac as noted identified the imaginary part with zitterbewegung effects and argued that this would be eliminated when averages over intervals of the order of the Compton scale are taken to recover meaningful physics [15]. Over the decades the significance of such cut off space time intervals has been stressed by T.D. Lee and several other scholars as noted earlier in Chapter 2 [130, 147, 150, 153]. Indeed we saw that with a minimum cut off length l, it was shown by Snyder that there would be a non commutative but Lorentz invariant spacetime structure. At the Compton scale we would have [16],

$$[x, y] = 0(l^2) \tag{5.35}$$

and similar relations.
In fact starting from the Dirac equation itself, we deduced directly the non commutativity (5.35) (Cf.refs.[158, 154]).
Let us now return to Dirac's formulation of the monopole in the light of the above comments. As noted above, the non integrability of the phase S in (5.32) gives rise to the electromagnetic field, while the nodal singularity gives rise to a term which is an integral multiple of 2π. As is well known [127] we have

$$\vec{\nabla} S = \vec{p} \tag{5.36}$$

where $\vec{p}$ is the momentum vector. When there is a nodal singularity, as noted above, the integral over a closed circuit of $\vec{p}$ does not vanish. In fact in this case we have a circulation given by

$$\Gamma = \oint \vec{\nabla} S \cdot d\vec{r} = \hbar \oint dS = 2\pi n \tag{5.37}$$

It is because of the nodal singularity that though the $\vec{p}$ field is irrotational, there is a vortex– –the singularity at the central point associated with the vortex makes the region multiply connected, or alternatively, in this region we cannot shrink a closed smooth curve about the point to that point. In fact if we use the fact as seen above that the Compton wavelength is a minimum cut off, then we get from (5.37) using (5.36), and on taking $n = 1$,

$$\oint \vec{\nabla} S \cdot d\vec{r} = \int \vec{p} \cdot d\vec{r} = 2\pi mc \frac{1}{2mc} = \frac{h}{2} \tag{5.38}$$

$l = \frac{\hbar}{2mc}$ is the radius of the circuit and $\hbar = 1$ in the above in natural units. In other words the nodal singularity or quantized vortex gives us the mysterious Quantum Mechanical spin half (and other higher spins for other values of n). In the case of the Quantum Mechanical spin, there are $2 \times n/2 + 1 = n + 1$ multiply connected regions, exactly as in the case of nodal singularities. Indeed in the case of the Dirac wave function, which is a bi-spinor $\begin{pmatrix} \Theta \\ \phi \end{pmatrix}$, as is known [181], far outside the Compton wavelength, it is the usual spinor Θ, preserving parity under reflections that predominates, whereas at and near the Compton scale it is the spinor

ϕ which predominates, where under a reflection ϕ goes over to $-\phi$.
To get a better insight, we consider a Hydrogen like atom in two dimensional space, for which the Schrodinger equation is given by [336, 319]

$$-\frac{\hbar^2}{2\mu}\left[\frac{1}{r}\frac{\partial}{\partial r}\left(r\frac{\partial}{\partial r}\right)+\frac{1}{r^2}\frac{\partial^2}{\partial\phi^2}\right]\psi(r,\phi)-\frac{Ze^2}{r}\phi(r,\phi)=E\psi(r,\phi) \quad (5.39)$$

As is well known the energy spectrum for (5.39) is given by

$$E=-\frac{Z^2e^4\mu}{2\hbar^2(n+m+\frac{1}{2})^2} \quad (5.40)$$

If we require that (5.40) be identical to the Bohr spectrum, then m should be a half integer, which also means that the configuration space is multiply connected. In the simplest case of a doubly connected space, we are dealing with $R^2 \times S^1$, where S^1 is a compactified space, generally considered to be a Kaluza-Klein space. However we would like to point out the following: The energy is given by

$$E=\left(k^2+\frac{S^2}{\rho^2}\right)\frac{\hbar^2}{2\mu} \quad (5.41)$$

In (5.41) there is an additional ground state energy $E = S^2\hbar^2/2\mu\rho^2$, where μ is the reduced mass and ρ is the radius of the compactified circle S^1. If ρ were to be the Planck length as in the Kaluza-Klein theory, then this extra energy becomes very large and is generally taken to be unobservable. On the other hand if ρ is taken to be the Compton wavelength as in our earlier discussion, then the above extra ground energy, as can be easily verified is of the same order as the usual energy.
In any case it can be seen that the Quantum Mechanical spin is a symptom of the multiply connected nature of Quantum spacetime, even in this non relativistic example. We remark that, as is well known, in (5.40), we can continue to take integral values of the quantum number m, provided, to the Coulomb potential energy an additional energy

$$\Delta E=\hbar\left(\sqrt{E/2\mu}\right)/r \quad (5.42)$$

is added. It is immediately seen that if in (5.42) r is of the order of the Compton wavelength, which is also $\sim e^2/mc^2$, then we recover e^2.To put it another way, if there was no Coulomb interaction in the conventional theory, then this additional contribution shows up as the Coulomb field. This points to the origin of the fundamental charge itself from topological conditions.

In Dirac's theory of displacement operators [15] as we saw, the operator $d_x \equiv \frac{d}{dx}$, if

$$0(\delta x^2) \neq 0 \tag{5.43}$$

becomes complex, and therefore, also the momentum operator, $p_x \equiv \imath\hbar d_x$ and the position operator. In other words if (5.43) holds good then we have to deal with complex or non-Hermitian coordinates. The implication of this is that (Cf.[299] for details) spacetime becomes non- commutative as seen above.

In any case here is the mysterious origin of the complex coordinates and spin. As noted it is the complex coordinates that lead from the Coulomb potential to the electromagnetic part of the Kerr-Newman metric and the electron's field including the anomalous gyro magnetic ratio which are symptomatic of the electron's spin [160]. It also means that the naked singularity is shielded by the fuzzy spacetime or equivalently the noncommutative geometry (5.35) (Cf. also [60]). Indeed, if we remember that $\vec{\nabla} S$ in (5.38) gives the momentum $\vec{p}$, we can see that (5.38) is an expression, indeed the origin, of the Wilson-Sommerfeld quantization rule [337, 318].

What all this means is that the presence of a Fermion in usual simply connected space tantamounts to making the space multiply connected−−like a hole in a sheet or like a vortex. In the case of the vortex, there are two velocities: that around the vortex, that is, the circulation, and that of the vortex itself. These correspond to the imaginary and real parts, respectively of the non-Hermitian (complex) momentum.

5.4 Lorentz Symmetry Violation Tests

Our fuzzy spacetime can be expected to lead to corrections to the Theory of Relativity, as the latter is based on the old ideas of differentiable spacetime whereas our approach yields a spacetime where there is non-commutativity and non-differentiability. Let us address this issue now.

Indeed we may already be observing a violation of Lorentz symmetry due to the observed time lags of cosmic gamma rays of different energies [290]. In fact it has been suspected that this could be the case from an observation of ultra high energy cosmic rays. In this case, given Lorentz symmetry there is the GZK cut off such that particles above an energy of about $10^{20}eV$ would not be able to travel cosmological distances and reach the earth (Cf. ref.[16, 338–343] for details). However, it is suspected that some twenty contra events have already been detected, and phenomenological models of

Lorentz symmetry violation have been constructed by Glashow, Coleman and others while this also follows from the author's fuzzy spacetime theory [344–349]. The essential point here is that the energy momentum relativistic formula is modified leading to a dispersive effect.

We would like to point out that apart from observation based models such a result follows from a fundamental point of view in modern approaches like ours in which the differentiable Minkowski spacetime is replaced by one which is fuzzy or noncommutative owing to a fundamental minimum length l being introduced.

Based on our earlier considerations, we can deduce from theory that the usual energy momentum formula is replaced by ($c = 1 = \hbar$) (Cf.[16, 338])

$$E^2 = m^2 + p^2 + \alpha l^2 p^4 \tag{5.44}$$

where α is a dimensionless constant of order unity. (For fermions, α is positive). To see this in greater detail, we note that, given a minimum length l, we saw that the usual commutation relations get modified and now become

$$[x, p] = \hbar' = \hbar[1 + \left(\frac{l}{\hbar}\right)^2 p^2]\, etc \tag{5.45}$$

where we have temporarily re-introduced $\hbar$ (Cf. also ref.[333]). (5.45) shows that effectively $\hbar$ is replaced by $\hbar'$. So,

$$E = [m^2 + p^2(1 + l^2 p^2)^{-2}]^{\frac{1}{2}}$$

or, the energy-momentum relation leading to the Klein-Gordon Hamiltonian is given by,

$$E^2 = m^2 + p^2 - 2l^2 p^4, \tag{5.46}$$

neglecting higher order terms.

For Fermions the analysis can be more detailed, in terms of Wilson lattices [350]. The free Hamiltonian now describes a collection of harmonic fermionic oscillators in momentum space. Assuming periodic boundary conditions in all three directions of a cube of dimension L^3, the allowed momentum components are

$$\mathbf{q} \equiv \left\{ q_k = \frac{2\pi}{L} \upsilon_k; k = 1, 2, 3 \right\}, \quad 0 \leq \upsilon_k \leq L - 1 \tag{5.47}$$

(5.47) finally leads to

$$E_{\mathbf{q}} = \pm \left(m^2 + \sum_{k=1}^{3} a^{-2} sin^2 q_k \right)^{1/2} \tag{5.48}$$

where $a = l$ is the length of the lattice, this being the desired result. (5.48) shows that α in (5.44) is positive. We have used the above analysis more to indicate that in the Fermionic case, the sign of α is positive. A rigid lattice structure imposes restrictions on the spacetime−−for example homogeneity and isotropy. Such restrictions are not demanded by fuzzy spacetime, and we use the lattice model more as a computational device (Cf. ref.[16]). This leads to a modification of the Dirac and Klein-Gordon equations at ultra high energies (Cf.ref.[16, 338, 339]). It may be remarked that proposals like equation (5.44) have been considered by several authors (Cf. refs.[351]-[360]). Our approach however, has been fundamental rather than phenomenological.

With this, it has been shown by the author that in the scattering of radiation, instead of the usual Compton formula we have

$$k = \frac{mk_0 + \alpha \frac{l^2}{2}[Q^2 + 2mQ]^2}{[m + k_0(1 - cos\Theta)]} \tag{5.49}$$

where we use natural units $c = \hbar = 1, m$ is the mass of the elementary particle causing the scattering, $\vec{k}, \vec{k_0}$ are the final and initial momentum vectors respectively and $Q = k_0 - k$, and Θ is the angle between the incident and scattered rays. Equation (5.49) shows that $k = k_0 + \epsilon$, where ϵ is a positive quantity less than or equal to $\sim l^2$, l being the fundamental length. It must be remembered that in these units k represents the frequency. The above can be written in more conventional form as

$$h\nu = h\nu_0[1 + 0(l^2)] \tag{5.50}$$

Equation (5.50) effectively means that due to the Lorentz symmetry violation in (5.44), the frequency is increased or, the speed of propagation of a given frequency is increased. As noted such models in a purely phenomenological context have been considered by Glashow, Coleman, Carroll and others. In any case what this means in an observational context is that higher frequency gamma rays should reach us earlier than lower frequency ones in the same burst. As Pavlopoulos reports (Cf.ref.[290]) this indeed seems to be the case.

Subject to further tests and confirmation, for example by NASA's GLAST satellite [361], spacetime at a microscopic length scale or ultra high energy level is not a smooth manifold, brought out by this manifestation in for example (5.44).

The small mass of the photon too, as seen in the last Chapter has consequences. It must be remembered that all these effects are small and consistent with the size of the Universe. Nevertheless there are experimental tests, in addition to those mentioned above, which are doable. As noted it is well known that for a massive vector field interacting with a magnetic dipole of moment $\mathbf{M}$, for example the earth itself, we would have with the usual notation (Cf.ref.[165])

$$\mathbf{A}(x) = \frac{\imath}{2} \int \frac{d^3k}{(2\pi)^3} \mathbf{M} \times \mathbf{k} \frac{e^{\imath k,x}}{\mathbf{k}^2 + \mu^2} = -\mathbf{M} \times \nabla \left(\frac{e^{-\mu r}}{8\pi r} \right)$$

$$\mathbf{B} = \frac{e^{-\mu r}}{8\pi r^3} |\mathbf{M}| \left\{ \left[\hat{r}(\hat{r} \cdot \hat{z}) - \frac{1}{3}\hat{z} \right] (\mu^2 r^2 + 3\mu r + 3) - \frac{2}{3}\hat{z}\mu^2 r^2 \right\} \tag{5.51}$$

We saw that considerations like this have yielded in the past an upper limit for the photon mass, for instance $10^{-48} gms$ in one estimate and $10^{-57} gms$ in later estimates. Nevertheless (5.51) can be used for a precise determination of the photon mass. It may be reiterated here that contrary to popular belief, there is no experimental evidence to indicate that the photon mass is zero! (Cf. discussion in ref.[301]).

Finally, we noted that the above value for the photon mass was also obtained by Terazawa [362], using the Dirac Large Number Hypothesis, something which is in fact a consequence of the Planck oscillator approach alluded to in Chapter 3 (Cf.ref.[16]).

5.5 The Finsler Spacetime Approach

In this approach not only the violation of GZK cut off in ultra high energy cosmic rays but also the anisotropy as indicated by data from COBE is taken into account. Then we have a Finsler metric [363–365], which in two dimensions can be written as

$$x' = e^{r\alpha} \quad L(x), tan\, \hbar\alpha = \frac{u}{c} \tag{5.52}$$

r being the anisotropy factor and L stands for the usual Lorentz transformation u being the velocity, which is not evident in (5.52) because there is only one space dimension. From observation it appears that

$$r \geq 10^{-10}$$

Finally the Finslerian metric in three dimensions is given by

$$ds^2 = \left[\frac{(dx^0 - dx^1)^2}{(dx^0)^2 - (dx^1)^2 - (dx^2)^2 - (dx^3)^2} \right]^r \cdot \left[(dx^0)^2 - (dx^1)^2 - (dx^2)^2 - (dx^3)^2 \right] \tag{5.53}$$

It can be seen from (5.53) that there is a prefered direction like the x^1 axis in this special choice. The metric in (5.53) leads to a modified energy momentum formula which is given by

$$\left[\frac{(E/c - \vec{p}\cdot\vec{\nu})^2}{E^2/c^2 - p^2}\right]^{-r} (E^2/c^2 - p^2) = m^2c^2(1-r)^{1-r}(1+r)^{1+r}(\nu^2 = 1) \tag{5.54}$$

In (5.54) the anisotropy direction is given by $\vec{\nu}$. As r is small, (5.54) simplifies to a form similar to (5.31), with suitable approximations.

5.6 Remarks

We would like to point out that a Lorentz symmetry violation would also imply a violation of the CPT invariance, though this could be expected only from high energies, the effect itself being small (Cf.ref.[366] and references therein). Indeed in the light of (5.44), the modified Dirac equation (Cf.ref.[339]) throws up a Lagrangian with a parity violating term. Specifically, this term in the Dirac Hamiltonian is

$$\gamma^5 l p^2,$$

which is clearly CPT violating.

It must also be remarked that given a fuzzy spacetime or equivalently a noncommutative geometry, we can deduce the photon mass. It has already been pointed out that such noncommutativity of coordinates leads to a term in gauge theory which is similar to the symmetry breaking Higgs field term [16]. It is this term which in the case of $U(1)$ electromagnetic field gives $\sim l^2$, which as argued is the photon mass [367]. From this point of view, the different routes to Lorentz symmetry violation, except the Finsler spacetime approach, are all really due to the $0(l^2)$ effect. However, as noted above, even the Finsler spacetime case, (5.54), approximates (5.44).

Finally, it must be pointed out that experiments by Mignani and coworkers [368–370] indicate a violation of Lorentz symmetry in the low energy regime while a more updated experimental set up for the photon mass effect in (5.51) is described in [284].

5.7 A Test for Non Commutative Spacetime

We first observe that non commutativity means that simple coordinates like x and y do not commute but rather we have relations of the type encountered earlier, viz.,

$$[x, y] = \beta\Theta \tag{5.55}$$

where $\beta \sim l^2, l$ being the minimum extension and Θ are matrices. The Equation (5.55) suggests that the coordinates also contain some type of a momentum with a suitable dimensional factor [371, 54]:

$$y = h' p_x \tag{5.56}$$

It has been shown that

$$h' = l^2/h \tag{5.57}$$

On the other hand, it has been shown by the author, Saito and others [297, 372] that the above non commutative geometry gives rise to a magnetic field $\vec{B}$ given by

$$Bl^2 = hc/e \tag{5.58}$$

where $B = |\vec{B}|$. Use of (5.58) with (5.56) and (5.57) gives

$$y = \frac{c}{eB} p_x \tag{5.59}$$

Equation (5.59) is familiar from the theory of a particle in a uniform magnetic field, first worked out by Landau [373, 308]. What happens there is, given a uniform magnetic field along the z axis, the particle in the classical sense executes circles in the x, y plane with quantized energy levels given by, with the usual notation

$$E = (n + \frac{1}{2}) h\omega_B + p_z^2/2m - \mu\sigma B/s \tag{5.60}$$

where ω_B is given by

$$\omega_B = |e|B/mc$$

while (5.59) holds good. These are the so called Landau levels. Thus the non commutative geometry would show up as a quantization of the energy as in (5.60). It must be observed that in (5.60) p_z is itself not quantized. It is surprising that Landau's original formulation of the particle in a uniform magnetic field also throws up a non commutative geometry––in fact the coordinates of the centre of the above concentric circles representing Landau levels do not commute. But this non commutative geometry was overlooked! Interestingly if one considers the usual Harmonic oscillator problem, but this time in the context of the non commutative geometry (5.55), then again we recover the Landau problem (Cf.ref.[374] for details).

Chapter 6

The Origin of Mass, Spin and Interaction

6.1 The Unification Mantra

If we look back at prehistory, we find bewildered man assigning to different natural phenomena, different controlling powers or deities. But gradually, we could discern underlying common denominators. Over the millennia man's quest for an understanding of the universe has been to perceive disparate phenomena in terms of a minimal set of simple principles. Today looking back we can see the logic of Occam's razor (literally, "A satisfactory proposition should contain no unnecessary complications"), or an economy of hypothesis– a far cry from prehistoric times.
In the words of F.J. Dyson[375], ".... the very greatest scientists in each discipline are unifiers. This is especially true in Physics. Newton and Einstein were supreme as unifiers. The great triumphs of Physics have been triumphs of unification. We almost take it for granted that the road of progress in Physics will be a wider and wider unification...".
Sir Isaac Newton was the first great unifier. He discovered the Universal Law of Gravitation: The force which kept the moon going round the earth, or the earth round the sun was also the force which kept binary stars going around each other and so on. All this was basically the same force of gravitation which brought apples down from a tree. This apart his Laws of Motion were also universal.
In the nineteenth century the work of Faraday, Ampere and others showed the close connection between the apparently totally dissimilar forces of electricity and magnetism. It was Maxwell who unified electricity not just with magnetism but with optics as well[376].
There was another great unification in the nineteenth century: Thermodynamics linked the study of heat to the kinetic theory of gases[377].

In the early part of the twentieth century Einstein fused space and time, giving them an inseparable identity, the Minkowski spacetime[378]. He went on to unify spacetime with Gravitation in his General Theory of Relativity[229]. However the unification of Electromagnetism and Gravitation has eluded several generations of physicists, Einstein included [39]. Meanwhile, thanks to the work of De Broglie and others, the newly born Quantum Theory unified the two apparently irreconcilable concepts of Newton's "particles" and Huygen's waves[15].
Yet another unification in the last century, which often is not recognised as such is the fusion of Quantum Mechanics and Special Relativity by Dirac, through his celebrated equation of the electron[15].
One more unification took place in the seventies due to the work of Salam, Weinberg, Glashow and others– the unification of Electromagnetism with the weak forces. This has given a new impetus to attempts for unifying all interactions, Gravitation included.
The weak force is one of two forces, the other being the strong force, discovered during the twentieth century itself. Earlier studies and work revealed that there seemed to be three basic particles in the Universe, the protons, the neutrons and the electrons. While the proton and the electron interact via the electromagnetic force, in the absence of this force the proton and the neutron appear to be a pair or a doublet. However the proton and the neutral neutron interact via "strong forces", forces which are about ten times stronger than the electromagnetic but have a much shorter range of just about $10^{-13}cms$. These are the forces which bind, for example, the protons in the nucleus.
The existence of the neutrino was postulated by Pauli in 1930 to explain the decay of the neutron, and it was discovered by Reines and Cowan in 1955. The weak force which is some 10^{-13} times the strength of the electromagnetic force is associated with neutrino type particles and has an even shorter range, $10^{-16}cms$. The neutrino itself has turned out to be one of the most enigmatic of particles, with peculiar characteristics, the most important of which is its handedness. This handedness property appears to be crucial for weak forces.
Later work revealed that while particles like the electron and neutrino, namely the leptons may be "truly" elementary, particles like the protons may be composite, infact made up of still smaller objects called quarks – six in all as we saw in Chapter 1 [35]. Today it is believed that the quarks interact via the strong forces or the QCD forces, we have already encountered.

All these "material" particles are Fermions, with half integral spin. Forces or interactions while originating in Fermions, are mediated by messengers like photons which are Bosons, with integral spin, spin 1 in fact. This is crucial, for, now there is the formalism of gauge theory which can describe all these interactions. We briefly saw all this.

In this sense Gravitation is not a gauge force. It is supposedly mediated by particles of spin 2.

To picturise the above let us consider the interaction between a proton and an electron. A proton could be imagined to emit a photon which is then absorbed by the electron. These studies, in the late forties and fifties culminated in the highly successful theory of Quantum Electro Dynamics or QED.

Instead of a single mediating particle we could think of multiplets, all having equal masses. With group theoretical inputs, one could shortlist, singlets with one particle like the photon, triplets, octets and so on as possible candidates[35].

Motivated by the analogy of Electromagnetism mediated by the spin one photon, it was realized in the fifties that the W^+ and W^- Bosons could be possible candidates for the mediation of the weak force. However there had to be one more messenger so that there would be the allowable triplet. It was suggested by Ward and Salam that the third candidate could be the photon itself, which would then provide not only a description of the weak force but would also unify it with Electromagnetism. However while the W particles were massive, the photon was massless so that they could not form a triplet. A heavy photon or Z^0 was then postulated to make up a triplet, while the photon was also used for the purpose of unification, and moreover a mixing of Z^0 and the photon was required for the well known renormalization, that is the removal of infinities.

The question was how could the photon be massless while the W and Z particles would be massive? It was suggested that this could be achieved through the spontaneous breaking of symmetry[35]. For example a bar magnet when heated, looses its magnetism. In effect the North and South pole symmetry is broken. Conversely, when the magnet cools down, polarity or asymmetry is restored spontaneously. This infact is a phase transition from symmetry to asymmetry.

In our case, before the spontaneous breaking of symmetry or the phase transition, the Ws, Zs, and the photons would all be massless. After the phase transition, while the photons remain mass less, the others would acquire mass. This phase transition would occur at temperatures $\sim 10^{15^\circ}$

Centigrade. At higher temperatures there would be a single electroweak force. As the temperature falls to the above level Electromagnetism and weak forces would separate out.
The next problem was, the inclusion of the strong forces. Clearly the direction to proceed appeared to be to identify the gauge character of the strong force– mediated by spin one particles, the gluons. (The approach differed from an earlier version of strong interaction in terms of Yukawa's pions.) This force binds the different quarks to produce the different elementary particles, other than the leptons. This is the standard model. It must be mentioned that in the standard model, the neutrino is a massless particle. However we have not yet conclusively achieved a unification of the electroweak force and the strong force. We proceed by the analogy of the electroweak unification to obtain a new gauge force that has been called by Jogesh Pati and Abdus Salam as the electro nuclear force, or in a similar scheme the Grand Unified Force by Glashow and Georgi.
It must be mentioned that one of the predictions is that the proton would decay with a life time of about 10^{32} years, very much more than the age of the Universe itself. However some believe that we are near a situation where this should be observable[379]. Others have given up on this idea. This "unifying" theory as we saw in Chapter 1, still relies on eighteen arbitrary parameters, apart from being plagued by problems like the "hierarchy problem", which arises from the widely different energies and therefore masses associated with the various interactions, the as yet non-existent monopole, infinities or divergences (which have to be eliminated by renormalization), and so on [380]-[382], [27].
The super Kamiokande determination of neutrino mass in the nineties, is the first evidence of what may be called, Physics beyond the standard model. Interestingly in this theory we would also require a right handed neutrino in this case.
Meanwhile extended particles had come into vogue from the seventies, with string theory[103, 383, 384]. Starting off with objects of the size of the Compton wavelength, the theory of superstrings now deals with the Planck length of about $10^{-33} cms$.
We have already noted that all interactions except Gravitation which is mediated by spin 2 gravitons are generalizations of the electromagnetic gauge theory. String theory combines Special Relativity, and General Relativity−−we need ten, $(9 + 1)$, dimensions for quantizing strings, and we also get a mass less particle of spin two which is the mediator of the gravitational force. This way there is the possibility of unifying all interac-

tions including Gravitation. Further, in the above ten dimensions there are no divergences. This is because the spatial extension of the string fudges the singularities (or vertices). However, we require, for verification of the string model, energies $\sim 10^{18} m_P$, as against the presently available $10^3 m_P$. For this and other reasons, as we briefly saw in Chapter 1, String theory too is falling out of favour.

As noted modern Fuzzy Spacetime and Quantum Gravity approaches to the problem deal with a non differentiable spacetime manifold. In the latter approach as we saw there is a minimum spacetime cut off, with, what is nowadays called a non commutative geometry, a feature shared by the Fuzzy Spacetime also [145, 315, 16, 316, 60, 317]. The new geometry is given by, as seen repeatedly,

$$[dx^\mu, dx^\nu] \approx \beta^{\mu\nu} l^2 \neq 0 \tag{6.1}$$

While equation (6.1) is true for any minimum cut off l, we will argue that it is most interesting and leads to physically meaningful relations including a rationale for the Dirac equation and the underlying Clifford algebra, when l is at the Compton scale (Cf.ref.[16]). In any case given (6.1), the usual invariant line element,

$$ds^2 = g_{\mu\nu} dx^\mu dx^\nu \tag{6.2}$$

has to be written in terms of the symmetric and nonsymmetric combinations for the product of the coordinate differentials. That is, the right side of equation (6.2) would become

$$\frac{1}{2} g_{\mu\nu} \left[(dx^\mu dx^\nu + dx^\nu dx^\mu) + (dx^\mu dx^\nu - dx^\nu dx^\mu) \right],$$

In effect we would have

$$g_{\mu\nu} = \eta_{\mu\nu} + k h_{\mu\nu} \tag{6.3}$$

So the noncommutative geometry introduces an extra term, that is the second term on the right side of (6.3). It has been shown in detail by the author that (6.1) or (6.2) lead to a reconciliation of electromagnetism and gravitation and lead to what may be called an extended gauge formulation [324, 371, 300, 385].

The extra term in (6.3) leads to an energy momentum like tensor but it must be stressed that its origin is in the non commutative geometry (6.1). All this of course is being considered at the Compton scale of an elementary particle.

6.2 Compton Scale Considerations

As in the case of General Relativity [229, 45], but this time remembering that neither the coordinates nor the derivatives commute we have

$$\partial_\lambda \partial^\lambda h^{\mu\nu} - (\partial_\lambda \partial^\nu h^{\mu\lambda} + \partial_\lambda \partial^\mu h^{\nu\lambda})$$

$$-\eta^{\mu\nu}\partial_\lambda \partial^\lambda h + \eta^{\mu\nu}\partial_\lambda \partial_\sigma h^{\lambda\sigma} = -kT^{\mu\nu} \tag{6.4}$$

It must be reiterated that the non commutativity of the space coordinates has thrown up the analogue of the energy momentum tensor of General Relativity, viz., $T^{\mu\nu}$. We identify this with the energy momentum tensor. At this stage, we note that the usual energy momentum tensor is symmetric, this being necessary for the conservation of angular momentum. This condition does not hold in (6.4), and the circumstance requires some discussion. Let us first consider the usual case with commuting coordinates [386]. Here as is well known, we start with the action integral

$$S = \int \Lambda \left(q, \frac{\partial q}{\partial x^\imath} \right) dV dt = \frac{1}{c} \int \Lambda d\Omega, \tag{6.5}$$

In (6.5) Λ is a function of the generalized coordinates q of the system, as also their first derivatives with respect to the space and time coordinates. In our case the q will represent the four potential A^μ (Cf.[386]) as will be seen again in (6.24). Requiring that (6.5) should be stationery leads to the usual Euler-Lagrange type equations,

$$\frac{\partial}{\partial x^\imath} \frac{\partial \Lambda}{\partial q, \imath} - \frac{\partial \Lambda}{\partial q} = 0 \tag{6.6}$$

In (6.6), the summation convention holds. We also have from first principles

$$\frac{\partial \Lambda}{\partial x^\imath} = \frac{\partial \Lambda}{\partial q} \frac{\partial q}{\partial x^\imath} + \frac{\partial \Lambda}{\partial q_{,k}} \frac{\partial q_{,k}}{\partial x^\imath} \tag{6.7}$$

At this stage we note that in the usual theory we have in (6.7)

$$\left\{ \frac{\partial q_{,k}}{\partial x^\imath} - \frac{\partial q_{,\imath}}{\partial x^k} \right\} = A^l_{k,\imath} - A^l_{\imath,k} = 0 \tag{6.8}$$

Using (6.8) it then follows that conservation of angular momentum requires

$$T^{\imath k} = T^{k\imath}, \tag{6.9}$$

However in our case the right side of (6.8) does not vanish due to the non commutativity of the coordinates and the partial derivatives, as will be seen more explicitly in (6.24). This means that the condition (6.9) does not hold

for non commutative coordinates, and hence (6.4) does not contradict the conservation of angular momentum. However there is new physics here and this new physics will be seen in equations following from (6.24): we recover Electromagnetism as an effect.
Remembering that $h_{\mu\nu}$ is a small effect, we can use the methods of linearized General Relativity [229, 45], to get from (6.4),

$$g_{\mu v} = \eta_{\mu v} + h_{\mu v}, h_{\mu v} = \int \frac{4T_{\mu v}(t - |\vec{x} - \vec{x}'|, \vec{x}')}{|\vec{x} - \vec{x}'|} d^3 x' \tag{6.10}$$

It was shown several years ago in the context of linearized General Relativity, that for distances $|\vec{x} - \vec{x'}|$ much greater than the distance $\vec{x'}$, that is well outside the Compton wavelength in our case, we can recover from (6.10) the electromagnetic potential (Cf.ref.[130] and references therein). We will briefly return to this point.
In (6.10) we use the well known expansions [45]

$$\bar{T}_{\mu\nu}(t - |x - x'|, x') = \sum_{n=0}^{\infty} \frac{1}{n} \left[\frac{\partial^n}{\partial t^n} \bar{T}_{\mu\nu}(t - r, x') \right] (r - |x - x'|)^n, \tag{6.11}$$

$$r - |x - x'| = x^j \left(\frac{x^{j'}}{r} \right) + \frac{1}{2} \frac{x^j x^k}{r} \left(\frac{x^{j'} x^{k'} - r'^2 \delta_{jk}}{r^2} \right) + \cdots, \tag{6.12}$$

$$\frac{1}{|x - x'|} = \frac{1}{r} + \frac{x^j}{r^2} \frac{x^{j'}}{r} + \frac{1}{2} \frac{x^j x^k}{r^3} \frac{(3x^{j'} x^{k'} - r'^2 \delta_{jk})}{r^2} + \cdots, \tag{6.13}$$

where $r \equiv |\vec{x}|$. We note that

$$r = |\vec{x}| \sim l \tag{6.14}$$

where l is of the order of the Compton wavelength. So the expansion of the integral in (6.10) now gives using (6.12) and (6.13),

$$\frac{T}{r} + T' \cdot \frac{1}{r}(r - |x - x'|) + \frac{1}{2} T'' \frac{(r - |x - x'|)}{r} \tag{6.15}$$

where primes denote the derivatives and we have dropped the superscripts for the moment. Denoting $(r - |x - x'|) \equiv r'$, where $0 \leq r' \leq r$, we can write

$$\langle r' \rangle \approx \gamma r \, \text{where} \, \gamma \sim 0(1) \tag{6.16}$$

Finally the expansion gives on the use of (6.15) and (6.16), the expression

$$\frac{T}{r} + \gamma T' + \frac{\gamma^2}{2} T'' \cdot r \tag{6.17}$$

That is we have, from (6.10) and (6.17),

$$h_{\mu v} = 4 \int \frac{T_{\mu v}(t, \vec{x}')}{|\vec{x} - \vec{x}'|} d^3 x' + (\text{terms independent of} \vec{x}) + 2$$
$$\int \frac{d^2}{dt^2} T_{\mu v}(t, \vec{x}') . |\vec{x} - \vec{x}'| d^3 x' + 0(|\vec{x} - \vec{x}'|^2) \quad (6.18)$$

The last term in (6.18) can be neglected, as we are dealing with points near the Compton wavelength. The first term gives on the use of (6.17), a Coulombic $\frac{\alpha}{r}$ type interaction except that the coefficient α is of much greater magnitude as compared to the gravitational or electromagnetic case, because in the expansion (6.12) and (6.13) all terms are of comparable order. The second term on the right side of (6.18) is of no dynamical significance as it is independent of $\vec{x}$. The third term however is of the form constant $\times r$. That is the potential (6.18) is exactly of the form of the QCD potential [30]

$$-\frac{\alpha}{r} + \beta r \quad (6.19)$$

In (6.19) α is of the order of the mass of the particle as follows from (6.18) and the fact that $T^{\mu \nu}$ is the energy momentum tensor given by

$$T^{\mu \nu} = \rho u^\mu u^\nu \quad (6.20)$$

where in (6.20), remembering that we are at the Compton scale, $u^\imath \sim c$. We now deduce two relations which can be deduced directly from the theory of the Dirac equation [15]. We do it here to show the continuity of the above theme. Remembering that from (6.1), we are within a sphere of radius given by the Compton length where the velocities equal that of light, as noted above, we have equations

$$|\frac{du_v}{dt}| = |u_v| \omega \quad (6.21)$$

$$\omega = \frac{|u_v|}{R} = \frac{2mc^2}{\hbar} \quad (6.22)$$

Alternatively as remarked, we can get (6.21) from the theory of the Dirac equation itself [15], viz.,

$$\imath \hbar \frac{d}{dt}(u_\imath) = -2mc^2(u_\imath),$$

Using (6.20), (6.21) and (6.22) we get

$$\frac{d^2}{dt^2} T^{\mu v} = 4 \rho u^\mu u^v \omega^2 = 4 \omega^2 T^{\mu v} \quad (6.23)$$

Equation (6.23) too is obtained in the Dirac theory (loc.cit). Whence, as can be easily verified, α and β in (6.19) have the correct values required for the QCD potential (Cf. also [130]). (Alternatively βr itself can be obtained, as in the usual theory by a comparison with the Regge angular momentum mass relation: It is in fact the constant string tension like potential mentioned in Chapter 1 which gives quark confinement and its value is as in the usual theory [387].)

Let us return to the considerations which lead via a non commutative geometry to an energy momentum tensor in (6.4). We can obtain from here the origin of mass and spin itself, for we have as is well known (Cf.ref.[45])

$$m = \int T^{00} d^3 x$$

and via

$$S_k = \int \epsilon_{klm} x^l T^{m0} d^3 x$$

the equation

$$S_k = c < x^l > \int \rho d^3 x \cdot$$

While m above can be immediately and consistently identified with the mass, the last equation gives the Quantum Mechanical spin if we remember that we are working at the Compton scale so that

$$\langle x^l \rangle = \frac{\hbar}{2mc} \cdot$$

Returning to the considerations in (6.1) to (6.4) it follows that (Cf.ref.[324])

$$\frac{\partial}{\partial x^\lambda} \frac{\partial}{\partial x^\mu} - \frac{\partial}{\partial x^\mu} \frac{\partial}{\partial x^\lambda} \text{goes over to} \frac{\partial}{\partial x^\lambda} \Gamma^\nu_{\mu\nu} - \frac{\partial}{\partial x^\mu} \Gamma^\nu_{\lambda\nu} \qquad (6.24)$$

Normally in conventional theory the right side of (6.24) would vanish. Let us designate this non vanishing part on the right by

$$\frac{e}{c\hbar} F^{\mu\lambda} \qquad (6.25)$$

We have shown here that the non commutativity in momentum components leads to an effect that can be identified with Electromagnetism and in fact from expression (6.25) we have

$$A^\mu = \hbar \Gamma^{\mu\nu}_\nu \qquad (6.26)$$

where $A_\mu \equiv q$, which we encountered in (6.6), as noted can be identified with the electromagnetic four potential and the Coulomb law deduced for

$|\vec{x}-\vec{x}'|$ in (6.10) much greater than $|\vec{x}'|$ that is well outside the Compton scale (Cf.ref.[16] and also ref. [130]). (Cf. also equation (6.3)). Indeed we have referred to this in the discussion after (6.4). It must be mentioned that despite non commutativity, we are using as an approximation the usual continuous partial derivatives, though these latter do not commute amongst themselves now. This facilitates the analysis and brings out the physical effects. In any case as can be seen from (6.1), the effects are of the order l^2.

To see this in the light of the usual gauge invariant minimum coupling (Cf.ref.[130]), we start with the effect of an infinitesimal parallel displacement of a vector in this non commutative geometry,

$$\delta a^\sigma = -\Gamma^\sigma_{\mu\nu} a^\mu dx^\nu \tag{6.27}$$

As is well known, (6.27) represents the effect due to the curvature and non integrable nature of space - in a flat space, the right side would vanish. Considering the partial derivatives with respect to the μth coordinate, this would mean that, due to (6.27)

$$\frac{\partial a^\sigma}{\partial x^\mu} \to \frac{\partial a^\sigma}{\partial x^\mu} - \Gamma^\sigma_{\mu\nu} a^\nu \tag{6.28}$$

Letting $a^\mu = \partial^\mu \phi$, we have, from (6.28)

$$D_{\mu\nu} \equiv \partial_\nu \partial^\mu \to D'_{\mu\nu} \equiv \partial_\nu \partial^\mu - \Gamma^\mu_{\lambda\nu}\partial^\lambda$$

$$= D_\mu - \Gamma^\mu_{\lambda\nu}\partial^\lambda \tag{6.29}$$

Now we can also write

$$D_{\mu\nu} = (\partial^\mu - \Gamma^\mu_{\lambda\lambda})(\partial_\nu - \Gamma^\lambda_{\lambda\nu}) + \partial^\mu \Gamma^\lambda_{\lambda\nu} + \Gamma^\mu_{\lambda\lambda}\partial_\nu$$

So we get

$$D_{\mu\nu} - \Gamma^\mu_{\lambda\lambda}\partial_\nu = (p^\mu)(p_\nu)$$

where

$$p^\mu \equiv \partial^\mu - \Gamma^\mu_{\lambda\lambda}$$

Or,

$$D_{\mu\mu} - \Gamma^\mu_{\lambda\lambda}\partial_\mu = (p^\mu)(p_\mu)$$

Further we have

$$D'_{\mu\mu} = D_{\mu\mu} - \Gamma^\mu_{\lambda\mu}\partial^\lambda$$

Thus, (6.29) gives, finally,

$$D'_{\mu\nu} = (p_\mu)(p_\nu)$$

That is we have

$$\frac{\partial}{\partial x^\mu} \to \frac{\partial}{\partial x^\mu} - \Gamma^\nu_{\mu\nu}$$

Comparison with (6.26) establishes the required identification.
It is quite remarkable that equation (6.26) is mathematically identical to Weyl's unification formulation, though as noted this was not originally acceptable because of the ad hoc insertion of the electromagnetic potential. Here in our case it is a consequence of the geometry - the noncommutative geometry.
We have also described how in the usual commutative spacetime the Dirac spinorial wave functions conceal the noncommutative character (6.1) [16]. Indeed we can verify all these considerations in a simple way as follows. To recapitulate, first let us consider the usual spacetime, in which the Dirac wave function is given by

$$\psi = \begin{pmatrix} \chi \\ \Theta \end{pmatrix},$$

where χ and Θ are two component spinors. It is well known that under reflection while the so called positive energy spinor Θ behaves normally, on the contrary $\chi \to -\chi, \chi$ being the so called negative energy spinor which comes into play at the Compton scale. That is, space is doubly connected. Because of this property as shown in detail [371], there is now a covariant derivative given by, in units, $\hbar = c = 1$,

$$\frac{\partial \chi}{\partial x^\mu} \to [\frac{\partial}{\partial x^\mu} - nA^\mu]\chi \tag{6.30}$$

where

$$A^\mu = \Gamma^{\mu\sigma}_\sigma = \frac{\partial}{\partial x^\mu} log(\sqrt{|g|}) \tag{6.31}$$

Γ denoting the Christofell symbols.
A^μ in (6.31)is now identified with the electromagnetic potential, exactly as in Weyl's theory except that now, A^μ arises from the bi spinorial character of the Dirac wave function or the double connectivity of spacetime. In other words, we return to (6.26) via an alternative (but connected) route.
What all this means is that the so called ad hoc feature in Weyl's unification theory is really symptomatic of the underlying noncommutative spacetime

geometry (6.1). Given (6.1) (or (6.3)) we get both Gravitation and Electromagnetism in a unified picture, because both are now the consequence of spacetime geometry. We could think that Gravitation arises from the symmetric part of the metric tensor (which indeed is the only term if $0(l^2)$ is neglected) and Electromagnetism from the antisymmetric part (which manifests itself as an $0(l^2)$ effect). It is also to be stressed that in this formulation, we are working with noncommutative effects at the Compton scale, this being true for the Weyl like formulation also.

6.3 Remarks

As we saw, the Compton scale comes as a Quantum Mechanical effect, within which we have zitterbewegung effects and a breakdown of causal Physics [15]. We, on the other hand have studied all this in the context of a non differentiable spacetime and noncommutative geometry.
Weinberg also noticed the non physical aspect of the Compton scale as we saw in detail [17]. Elaborating on the non-causal behavior he goes on:
"There is only one known way out of this paradox. The second observer must see a particle emitted at x_2 and absorbed at x_1. But in general the particle seen by the second observer will then necessarily be different from that seen by the first. For instance, if the first observer sees a proton turn into a neutron and a positive pi-meson at x_1 and then sees the pi-meson and some other neutron turn into a proton at x_2, then the second observer must see the neutron at x_2 turn into a proton and a particle of negative charge, which is then absorbed by a proton at x_1 that turns into a neutron. Since mass is a Lorentz invariant, the mass of the negative particle seen by the second observer will be equal to that of the positive pi-meson seen by the first observer. There is such a particle, called a negative pi-meson, and it does indeed have the same mass as the positive pi-meson. This reasoning leads us to the conclusion that for every type of charged particle there is an oppositely charged particle of equal mass, called its antiparticle. Note that this conclusion does not obtain in non-relativistic quantum mechanics or in relativistic classical mechanics; it is only in relativistic quantum mechanics that antiparticles are a necessity. And it is the existence of antiparticles that leads to the characteristic feature of relativistic quantum dynamics, that given enough energy we can create arbitrary numbers of particles and their antiparticles."
We reiterate however that in Weinberg's analysis, one observer sees only

protons at x_1 and x_2, whereas the other observer sees only neutrons at x_1 and x_2 while in between, the first observer sees a positively charged pion and the second observer a negatively charged pion. In a sense this is another perspective on the charge independence of strong interactions (or Heisenberg's isospin). We remark that in Weinberg's explanation which is in the spirit of the Feynman-Stuckleberg diagrams there is no charge conservation, though the Baryon number is conserved. The explanation for this is to be found in the considerations leading from (6.10) to (6.19)——within the Compton scale we have the QCD interactions——electromagnetic interaction is outside the Compton scale.

Our analysis uses the Compton length (and time) as the fundamental parameter. It may be added that there is a close parallel between the above considerations and the original Dirac monopole theory: in the latter it is the nodal singularity that gives rise to magnetism, while in the former, the multiply connected nature of space (or non commutativity) gives rise to Electromagnetism. This has been discussed in [297]and we will return to it. So too, it may be mentioned that the considerations in equations (6.21), (6.22) and (6.23) are connected with Dirac's membrane (and more recently and generally the p-brane) theory [367] - though Dirac himself approached the membrane problem from a different route. We will shortly come to this point.

Finally, it may be pointed out that Einstein himself always disliked the energy momentum tensor in his General Relativistic equation [388] as it was mechanical and non geometric! Pleasingly, in the above formulation, this term has a geometric origin——albeit, a non commutative geometry which also provides a unified description of linearized General Relativity and Quantum Mechanics.

6.4 Fuzzy Spacetime and Fermions

We now address the question: Can we take an alternative route to use Bosonic Strings which are at the real world Compton scale to obtain a description of Fermions without going to the Planck scale? We have already seen that Bosonic particles could be described as extended objects at the Compton scale. Let us rewrite, following Snyder, the following Lorentz invariant relations,

$$[x, y] = (\imath a^2/\hbar)L_z, [t, x] = (\imath a^2/\hbar c)M_x, etc.$$

$$[x, p_x] = \imath\hbar[1 + (a/\hbar)^2 p_x^2]; \cdots \tag{6.32}$$

If a^2 in (6.32) is neglected, then we get back the usual canonical commutation relations of Quantum Mechanics. This limit to an established theory is another attractive feature of (6.32).
However if order of a^2 is retained then the first of equations (6.32) as we have repeatedly seen, characterize a completely different spacetime geometry, one in which the coordinates do not commute. This is a noncommutative geometry and indicates that spacetime within the scale defined by a is ill defined, or is fuzzy [60]. Indeed in M-Theory too, we have a noncommutative geometry like ((6.32)). As we started with a minimum extention at the Compton scale, let us take $a = (l, \tau)$.
We also saw this by starting from the usual Dirac coordinate [15]

$$x_\imath = \left(c^2 p_\imath H^{-1} t\right) + \frac{1}{2} c\hbar \left(\alpha_\imath - c p_\imath H^{-1}\right) H^{-1} \tag{6.33}$$

where the α's are given by

$$\vec{\alpha} = \begin{bmatrix} \vec{\sigma} & 0 \\ 0 & \vec{\sigma} \end{bmatrix} \quad , \tag{6.34}$$

the σ's being the usual Pauli matrices. The first term on the right side of (6.33) is the usual Hermitian position coordinate. It is the second or imaginary term which contains $\vec{\alpha}$ that makes the Dirac coordinate non Hermitian. However we can easily verify from the commutation relations of $\vec{\alpha}$, using (6.34) that

$$[x_\imath, x_j] = \beta_{\imath j} \cdot l^2 \tag{6.35}$$

In fact (6.35) is just a form of the first of equations (6.32) and brings out the fuzzyness of spacetime in intervals where order of l^2 is not neglected.
We now obtain a rationale for the Dirac equation and spin from (6.35) [389, 371]. Under a time elapse transformation of the wave function, (or, alternatively, as a small scale transformation),

$$|\psi' >= U(R)|\psi > \tag{6.36}$$

we get

$$\psi'(x_j) = [1 + \imath\epsilon(\imath x_j \frac{\partial}{\partial x_j}) + 0(\epsilon^2)]\psi(x_j) \tag{6.37}$$

Equation (6.37) can be shown to lead to the Dirac equation when ϵ is the Compton time. A quick way to see this is as follows: At the Compton scale we have,

$$|\vec{L}| = |\vec{r} \times \vec{p}| = |\frac{\hbar}{2mc} \cdot mc| = \frac{\hbar}{2},$$

that is, at the Compton scale we get the Quantum Mechanical spin from the usual angular momentum. Next, we can easily verify, that the choice,

$$t = \begin{pmatrix} 1 & 0 \\ 0 & -1 \end{pmatrix}, \vec{x} = \begin{pmatrix} 0 & \vec{\sigma} \\ \vec{\sigma} & 0 \end{pmatrix} \tag{6.38}$$

provides a representation for the coordinates in (6.32), apart from scalar factors. As can be seen, this is also a representation of the Dirac matrices. Substitution of the above in (6.37) leads to the Dirac equation

$$(\gamma^\mu p_\mu - mc^2)\psi = 0$$

because

$$E\psi = \frac{1}{\epsilon}\{\psi'(x_j) - \psi(x_j)\}, \quad E = mc^2,$$

where $\epsilon = \tau$ (Cf.ref.[152]).

All this is symptomatic of an underlying fuzzy spacetime described by a noncommutative space time geometry (6.35) or (6.32) [154].

The point here is that under equation (6.35) and (6.38), the coordinates $x^\mu \to \gamma^{(\mu)} x^{(\mu)}$ where the brackets with the superscript denote the fact that there is no summation over the indices. In fact, in the theory of the Dirac equation it is well known [390]that,

$$\gamma^k \gamma^l + \gamma^l \gamma^k = -2g^{kl} I \tag{6.39}$$

where γ's satisfy the usual Clifford algebra of the Dirac matrices, and can be represented by

$$\gamma^k = \sqrt{2} \begin{pmatrix} 0 & \sigma^k \\ \sigma^{k*} & 0 \end{pmatrix} \tag{6.40}$$

where σ's are the Pauli matrices. Bade and Jehle noted that (Cf.ref.[390]), we could take the σ's or γ's in (6.40) and (6.39) as the components of a contravariant world vector, or equivalently we could take them to be fixed matrices, and to maintain covariance, to attribute new transformation properties to the wave function, which now becomes a spinor (or bi-spinor). This latter has been the traditional route, because of which the Dirac wave function has its bi-spinorial character. In this latter case, the coordinates retain their usual commutative or point character. It is only when we consider the equivalent former alternative, that we return to the noncommutative geometry (6.35).

That is, in the usual commutative spacetime the Dirac spinorial wave functions conceal the noncommutative character (6.35).

6.5 Branes

The considerations leading from (6.35) to (6.40) show that we are essentially dealing with a Clifford or C-space [391]. We will study this briefly, following [391]. Given the γ matrices which we encountered earlier we can write

$$\gamma_\mu \cdot \gamma_\nu \equiv \frac{1}{2}(\gamma_\mu \gamma_\nu + \gamma_\nu \gamma_\mu) = g_{\mu\nu}. \tag{6.41}$$

$$\gamma_\mu \wedge \gamma_\nu = \frac{1}{2}(\gamma_\mu \gamma_\nu - \gamma_\nu \gamma_\mu) \equiv \frac{1}{2}[\gamma_\mu, \gamma_\nu]. \tag{6.42}$$

In other words (6.42) gives an antisymmetrical tensor.
More generally we can consider a complete set of basis vectors γ_μ in a n-dimensional space satisfying (6.41) and (6.42). We can then have

$$\gamma_{\mu_1} \wedge \gamma_{\mu_2} \wedge \gamma_{\mu_3} = \frac{1}{3!}[\gamma_{\mu_1}, \gamma_{\mu_2}, \gamma_{\mu_3}], \tag{6.43}$$

$$\vdots \tag{6.44}$$

$$\gamma_{\mu_1} \wedge \gamma_{\mu_2} \wedge \cdots \wedge \gamma_{\mu_n} = \frac{1}{r!}[\gamma_{\mu_1}, \gamma_{\mu_2}, \cdots, \gamma_{\mu_n}]. \tag{6.45}$$

The left sides of (6.43), (6.44) and (6.45) are termed p-vectors, where p takes on the values, $3, 4, \cdots n$. A point in this n-dimensional space can be designated as in the previous section by

$$x = x^\mu \gamma_\mu. \tag{6.46}$$

More generally we have poly vectors which are obtained by superposing multivectors as follows

$$X = \sigma 1 + x^\mu \gamma_\mu + \frac{1}{2} x^{\mu_1 \mu_2} \gamma_{\mu_1 \mu_1} + \cdots + \frac{1}{n!} x^{\mu_1 \cdots \mu_n} \gamma_{\mu_1 \cdots \mu_n} \equiv x^M \gamma_M. \tag{6.47}$$

where

$$\gamma_{\mu_1 \cdots \mu_r} \equiv \gamma_{\mu_1} \wedge \gamma_{\mu_2} \wedge \cdots \wedge \gamma_{\mu_r}$$

and

$$x^M = (\sigma, x^\mu, x^{\mu_1 \mu_2}, \cdots, x^{\mu_1 \cdots \mu_r}),$$

$$\gamma_M = (1, \gamma_\mu, \gamma_{\mu_1 \mu_2, \cdots,} \gamma_{\mu_1 \cdots \mu_r}), \quad \mu_1 < \mu_2 < \cdots < \mu_r \tag{6.48}$$

The coordinate $X_{\mu_1 \cdots \mu_p}$ is a p-area enclosed by a loop of dimension $p - 1$. We now observe that the coordinates $\sigma, x_\mu, x_{\mu_1 \mu_2}$ etc. describe extended objects and that x^M is a quantity that assumes any real value and that

all possible X forms constitute a 2^n dimensional manifold, which we call the C-space. It may be mentioned that such higher dimensional extended objects or surfaces- the D-branes, were introduced by Polchinski [392].
In any case we can see that the C-space generates branes of different dimensionality as in M-Theory. If we stop with x_μ, we have the point space time of Bosons, if we stop with $x^\mu \gamma_\mu$ (in (6.47)), then we have the Fermions (of the earlier section) and finally we get branes by retaining other terms in (6.47). Moreover as we saw, retaining the usual coordinates x^μ tantamounts to neglecting $O(l^2)$, while for Fermions we retain those terms which are $\sim 10^{-22} cm^2$ in our Compton wavelength description, while if we retain term $O(l^3)$ for example, these are $\sim 10^{-33} cm^3$ (the Planck scale) and so on. However, it should be noted that we are really dealing with areas, volumes etc. in these higher order terms, and fractal dimensions as these are resolution dependent.

6.6 Dirac's Membrane and p-branes

In 1962 Dirac introduced a model of the electron which in its simplest terms was a spherical shell [51]. The important features of this model were that the electron had a finite self energy and only two parameters were required, viz., the mass and charge, as in the point particle case. There were excited states describing possibly the spectrum of heavier particles. On the other hand, Dirac's action did not contain the minimum coupling terms between the charge and the electromagnetic field. This coupling was obtained by a boundary condition and was consistent in the special gauge in which the potential on the membrane's surface was zero. Later this model was studied by Barut, Pavsic and others [7, 393, 394]. In these studies a covariant theory of a moving charge membrane in an arbitrary dimension coupled to the electromagnetic field was considered and developed. Interestingly there has been a return to similar ideas in M-theory, which is currently in vogue amongst superstring theorists. In general p-branes are being considered. In M-theory as we saw, coordinates become matrices and this leads to a noncommutative geometry [16].
We would like to point out that the above brane prescription can be obtained in a straightforward manner, originating from the original Dirac theory of the electron itself, something which has been long overlooked.
Indeed in the theory of the Dirac equation for the electron [15] we have

effectively, as noted a little earlier,

$$\imath\hbar\frac{d}{dt}(u_\imath) = -2mc^2(u_\imath), \tag{6.49}$$

$$\imath\hbar\frac{d^2}{dt^2}(u_\imath) = 2mc^2(\dot{u}_\imath), \tag{6.50}$$

(Dirac himself used the notation $c\alpha_\imath$ for $u_\imath$).

We would like to point out that these equations imply that the electron is a rotating shell at the Compton wavelength we described in equations (6.21) and (6.22). For in this case we would have (Cf.ref.[130] for details)

$$|\frac{du_\nu}{dt}| = |u_\nu|\omega,$$

where,

$$\omega = \frac{|u_\nu|}{R} = \frac{2mc^2}{\hbar}$$

These can be compared with (6.49) and (6.50) above. These equations would also follow directly from our earlier noncommutative geometry viz.,

$$[dx^\mu, dx^\nu] \approx \beta^{\mu\nu}l^2 \neq 0 \tag{6.51}$$

if l were at the Compton scale.

On the other hand we have argued that the noncommutative geometry (6.51) which is valid if there is a minimum space time cut off at an arbitrary length l, is particularly interesting, when l is the Compton wavelength.

So the ingredients for the shell model and p-branes of M-theory were already present but overlooked in Dirac's original electron theory once a noncommutative geometry or equivalently a minimum cut off at the Compton scale is considered [367].

6.7 A Modified Klein-Gordan Equation

Owing to the modified dispersion relation considered in the previous Chapter we have,

$$(D + l^2\nabla^4 - m^2)\psi = 0 \tag{6.52}$$

where D denotes the usual D'Alembertian.

Just to get a feel, it would be interesting to consider the extra effect in (6.52). For simplicity we take the one dimensional case. As in conventional theory if we separate the space and time parts of the wave function we get

$$l^2u^{(4)} + u^{(2)} + \lambda u = 0, \quad \lambda = E^2 - m^2 > 0 \tag{6.53}$$

where $u^{(n)}$ denotes the nth space derivative.
Whence if in (6.53) we take,

$$u = e^{\alpha x}$$

and $\alpha^2 = \beta$ we get,

$$l^2\beta^2 + \beta + \lambda = 0$$

whence

$$\beta = \frac{-1 \pm \sqrt{1 - 4l^2\lambda}}{2l^2}$$

So

$$\beta \approx \frac{-1 \pm \{1 - 2l^2\lambda\}}{2l^2} \tag{6.54}$$

From (6.54) it is easy to deduce that there are two extra solutions, as can be anticipated by the fact that (6.52) is a fourth order equation, unlike the usual second order Klein-Gordan equation. Thus we have

$$\beta = -\lambda (< 0)$$

giving the usual solutions, but additionally we have

$$\beta = -\left(\frac{1 - \lambda l^2}{l^2}\right) (< 0) \tag{6.55}$$

What do the two extra solutions in (6.55) indicate? To see this we observe that α is given by, from (6.55)

$$|\alpha| \approx \pm \frac{1}{l} \tag{6.56}$$

In other words (6.56) corresponds to waves with wavelength of the order l, which is intuitively quite reasonable.
What is interesting is that if l is an absolute length then the extra effect is independent of the mass of the particle. In any case the solutions from (6.56) are GZK violating solutions, arising as they do, from the modified energy momentum formula of the previous Chapter.
We now make some remarks. Departures from Lorentz symmetry of the type seen have as noted, been studied, though from a phenomenological point of view [344–349]. These arise mostly from an observation of Ultra High Energy Cosmic Rays. Given Lorentz Symmetry, there is the GZK cut off already alluded to, such that particles above this cut off would not be able to travel cosmological distances and reach the earth. However as mentioned, there are indications of a violation of the GZK cut off (Cf.references

[344]-[349]).
In any case some of the effects can be detected, it is hoped by the GLAST Satellite [361].
Interestingly, if in (6.52) we take, $-l^2$ rather than $+l^2$, we get two real exponential solution of (6.52). One of them is an increasing exponential leading to very high probabilities for finding these particles.

6.8 A Modified Dirac Equation

Once we consider a discrete spacetime structure, the energy momentum relation, as noted in the previous Chapter, gets modified [130, 350] and we have in units $c = 1 = \hbar$,

$$E^2 - p^2 - m^2 + l^2 p^4 = 0 \tag{6.57}$$

l being a minimum length interval, which could be the Planck length or more generally the Compton length. Let us now consider the Dirac equation

$$\{\gamma^\mu p_\mu - m\} \psi \equiv \{\gamma^\circ p^\circ + \Gamma\} \psi = 0 \tag{6.58}$$

If we include the extra effect shown in (6.57) we get

$$\left(\gamma^\circ p^\circ + \Gamma + \beta l p^2\right) \psi = 0 \tag{6.59}$$

β being a suitable matrix.
Multiplying (6.59) by the operator

$$\left(\gamma^\circ p^\circ - \Gamma - \beta l p^2\right)$$

on the left we get

$$p_0^2 - \left(\Gamma\Gamma + \{\Gamma\beta + \beta\Gamma\} + \beta^2 l^2 p^4\right\} \psi = 0 \tag{6.60}$$

If (6.60), as in the usual theory, has to represent (6.57), then we require that the matrix β satisfy

$$\Gamma\beta + \beta\Gamma = 0, \quad \beta^2 = 1 \tag{6.61}$$

From the properties of the Dirac matrices [181] it follows that (6.61) is satisfied if

$$\beta = \gamma^5 \tag{6.62}$$

Using (6.62) in (6.59), the modified Dirac equation finally becomes

$$\{\gamma^\circ p^\circ + \Gamma + \gamma^5 l p^2\} \psi = 0 \tag{6.63}$$

Owing to the fact that we have [181]

$$P\gamma^5 = -\gamma^5 P \tag{6.64}$$

It follows that the modified Dirac equation (6.63) is not invariant under reflections. This is a result which is to be expected because the correction to the usual energy momentum relation, as shown in (6.57) arises when l is of the order of the Compton wavelength. The usual Dirac four spinor $\begin{pmatrix} \Theta \\ \chi \end{pmatrix}$ as seen has the so called positive energy (or large) components Θ and the negative energy (or small) components χ. However, when we approach the Compton wavelength, that is as

$$p \to mc$$

the roles are reversed and it is the χ components which predominate. Moreover the χ two spinor as noted, behaves under reflection as [181]

$$\chi \to -\chi$$

In any case, as noted in the previous Chapter, this too provides an experimental test. We can also see that due to the modified Dirac equation (6.63), there is no additional effect on the anomalous gyromagnetic ratio. This is because, in the usual equation from which the magnetic moment is determined [395] viz.,

$$\frac{d\vec{S}}{dt} = -\frac{e}{\mu c}\vec{B} \times \vec{S},$$

where $\vec{S} = \hbar \sum /2$ is the electron spin operator, there is now an extra term

$$\left[\gamma^5, \sum\right] \tag{6.65}$$

However the expression (6.65) vanishes by the property of the Dirac matrices.

We remark that it has already been argued in detail that [129, 130] as we approach the Compton wavelength, the Dirac equation describes the quark with the fractional charge and handedness. Our above derivation and conclusion is pleasingly in agreement with this result.

Chapter 7

The Enigma of Gravitation

7.1 Gravitation in a New Light

More than five thousand years ago, the Rig Veda repeatedly raised the question: "How is it that *though unbound* the sun does not fall down?"
This was a question that puzzled thinking men over the millennia. Indian scholars right up to Bhaskaracharya who lived about a thousand years ago, believed in some attractive force which was responsible for keeping the celestial bodies from falling down.
The same problem as we saw in Chapter 1, was addressed by Greek thinkers about two thousand five hundred years ago. They devised transparent material spheres to which each of the celestial objects was attached−−the material spheres prevented them from falling down. Further, all motions were circular, for, the Greeks believed, taking the cue from Plato, that circles and spheres were perfect figures. The word orbit, which comes from orb, Greek for circle, is a vestige of that legacy.
Unfortunately too, it was this answer to the age old question, which held up further scientific progress till the time of Kepler, for even Copernicus accepted the transparent material spheres.
Kepler as we saw, had a powerful tool in the form of the accurate observations of Tycho Brahe. He also had the advantage of the Indian numeral system, which via the Arabs had reached Europe just a few centuries earlier. These lead him to his famous laws of elliptical orbits with definite periods correlated to distances from the Sun. This couching of natural phenomena in the terse language of mathematical symbols that could be manipulated, was the beginning of modern science.
The important point was that the Greek answer to the problem of why heavenly objects do not fall down−−the transparent material spheres which

held them up−−was now demolished. The age old question of why celestial bodies do not fall down came back to haunt again. Kepler himself speculated about some type of a magnetic force between the Sun and the planets, rather on the lines of earlier speculations in India.

It was Newton who provided the breakthrough.

To quote Hawking [396], "*The Philosophiae Naturalis Principia Mathematica* by Isaac Newton, first published in Latin in 1687, is probably the most important single work ever published in the physical sciences. Its significance is equalled in the biological sciences only by *The Origin of Species.* The original impulse which caused Newton to write the *Principia* was a question from Edmund Halley as to whether the elliptical orbits of the planets could be accounted for on the hypothesis of an inverse square force directed towards the Sun. This was something that Newton had worked out some years earlier but had not published, like most of his work on mathematics and physics. However, Halley's challenge, and the desire to refute the suggestions of others such as Hooke and Descartes, spurred Newton to try to write a proper account of this result."

Newton using Galileo's ideas of Mechanics, thus stumbled upon the Universal Law of Gravitation. It was audacious to dub the law universal−−for it was observed only for the solar system. But Herschel, in the next century noticed that the orbits of binary stars too, followed the law−−so it was truly universal.

This held sway for nearly two hundred and twenty five years, before Einstein came out with his own theory of Gravitation. There was no force in the mechanical sense that Newton and preceding scholars had envisaged it to be. Rather it was due to the curvature of spacetime itself. Einstein's bizarre ideas have had some experimental verification as we saw briefly in Chapter 3, while there are some other experimental consequences, such as gravitational waves, which need to be confirmed.

After Einstein's formulation of Gravitation a problem that has challenged and defied solution as we have seen, has been that of providing a unified description of Gravitation along with other fundamental interactions. Infact Einstein spent the last decades of his life in this fruitless quest. As he would lament [397] "I have become a lonely chap who is mainly known because he doesn't wear socks and who is exhibited as a curiosity on special occasions."

One of the earliest attempts was as seen earlier that of Hermann Weyl, which though elegant was rejected on the grounds that in the final analysis, it was not really a unification of Gravitation with Electromagnetism but

rather an adhoc prescription. However his original Gauge Invariant Geometry lead to the modern Gauge Theory for other interactions. We saw that within the framework of our theory, it is possible to get an extended gauge theory that includes Gravitation. A new generation of efforts to quantize the gravitational field, notably those of de Witt and later workers [398] have also not been successful: the gauge approach to elementary particle interactions has not lead to fruitful results in Gravitation. As Witten observed [243] "the existence of gravity clashes with our description of the rest of physics by quantum fields". An important reason is obvious: Other fields are described by vector bosons (with spin one), but gravitons are spin 2 bosons (or the gravitational field is a tensor field).
Modern approaches to this problem have as discussed, finally lead to the abandonment of a smooth spacetime manifold. Instead, the Planck scale is now taken to be a minimum fundamental scale. Let us revisit some of our earlier ideas, already encountered.
We have argued from different points of view to arrive at the otherwise empirically known equations [157, 158, 175]

$$R = \sqrt{\bar{N}} l_P = \sqrt{N} l$$

$$l = \sqrt{n} l_P \tag{7.1}$$

where l_P, l and R are the Planck length, the pion Compton wavelength and the radius of the universe and $N, \bar{N}$ and n are certain large numbers. Some of these are well known empirically for example $N \sim 10^{80}$ being the number of elementary particles, which typically are taken to be pions in the literature, in the Universe as we noted.
One way of arriving at the above relations as we saw is by considering a series of N Planck mass oscillators which are created out of the Quantum vacuum. In this case (Cf. also ref.[273]) we have

$$r = \sqrt{N a^2} \tag{7.2}$$

In (7.2) a is the distance between the oscillators and r is the extent. Equations (7.1) follow from equation (7.2).
We would like to point out that there is another way of arriving at equations (7.1) (Cf.ref.[175]). For this, we observe that the position operator for the Klein-Gordan equation is given by [399],

$$\vec{X}_{op} = \vec{x}_{op} - \frac{\imath \hbar c^2}{2} \frac{\vec{p}}{E^2}$$

Whence we get

$$\hat{X}_{op}^2 \equiv \frac{2m^3c^4}{\hbar^2} X_{op}^2 - \alpha = \frac{2m^3c^6}{\hbar^2} x^2 + \frac{p^2}{2m} \quad (7.3)$$

where α is a constant scalar, irrelevant in further discussion.
It can be seen that purely mathematically (7.3) for $\hat{X}_{op}^2$ defines the Harmonic oscillator equation, this time with quantized, what may be called space levels. It turns out that these levels are all multiples of $(\frac{\hbar}{mc})^2$. This Compton length is the Planck length for a Planck mass particle. Accordingly we have for any system of extension r,

$$r^2 \sim Nl^2$$

which gives back equation (7.1). This should not be surprising, because the Klein-Gordon equation describes a string of normal mode oscillators. We have noted that the Planck length is also the Schwarzschild radius of a Planck mass, that is we have

$$l_P = Gm_P/c^2 \quad (7.4)$$

Using equations (7.1) and (7.4), we will now re-derive a few new and valid and a number of otherwise empirically known relations involving the various microphysical parameters and large scale parameters. Some of these relations are deducible from the others. Many of these relations featured (empirically) in Dirac's Large Number Cosmology. We follow Dirac and Melnikov in considering $l, m, \hbar, l_P, m_P$ and e as microphysical parameters [43, 201]. Large scale parameters include the radius and the mass of the universe, the number of elementary particles in the universe and so on.
In the process we will also examine the nature of gravitation. It must also be observed that the Large Number relations below are to be considered in the Dirac sense, wherein for example the difference between the electron and pion (or proton) masses is irrelevant [17].
We have used the following well known equation which we obtained through different routes:

$$\frac{GM}{c^2} = R \quad (7.5)$$

For example in an uniformly expanding flat Friedman spacetime, we have [17]

$$\dot{R}^2 = \frac{8\pi G\rho R^2}{3}$$

If we substitute $\dot{R} = c$ at the radius of the universe in the above we recover (7.5).

We now observe that from the first two relations of (7.1), using the Compton wavelength expression we get

$$m = m_P / \sqrt{n} \tag{7.6}$$

Using also the second relation in (7.1) we can easily deduce

$$\bar{N} = Nn \tag{7.7}$$

Using (7.1) and (7.5) we have

$$M = \sqrt{\bar{N}} m_P \tag{7.8}$$

Interestingly (7.8) can be obtained directly, without recourse to (7.5), from the energy of the Planck oscillators (Cf.ref.[158]). Combining (7.8) and (7.6) we get

$$M = \left(\sqrt{\bar{N}n}\right) m \tag{7.9}$$

Further if we use in the last of equation (7.1) the fact that l_P is the Schwarzchild radius, that is equation (7.4), we get,

$$G = \frac{lc^2}{nm} \tag{7.10}$$

We now observe that if we consider the gravitational energy of the $\bar{N}$ Planck masses (which do not have any other interactions) we get,

$$\text{Gravitational Energy} = \frac{G\bar{N}m_P^2}{R}$$

If this is equated to the inertial energy in the universe, Mc^2, as can be easily verified we get back (7.5). In other words the inertial energy content of the universe equals the gravitational energy of all the $\bar{N}$ Planck oscillators. This is yet another derivation of (7.5).

Similarly if we equate the gravitational energy of the n Planck oscillators constituting the pion we get

$$\frac{Gm_P^2 n}{R} = mc^2 \tag{7.11}$$

Using in (7.11) equation (7.4) we get

$$\frac{l_P m_P n}{R} = m$$

Whence it follows on using (7.7), (7.6) and (7.1),

$$n^{3/2} = \sqrt{\bar{N}}, \, n = \sqrt{N} \tag{7.12}$$

Substituting the value for n from (7.12) into (7.10) we will get

$$G = \frac{lc^2}{\sqrt{N}m} \tag{7.13}$$

a relation we deduced in Chapter 3, alternatively. If we use (7.12) in (7.9) we will get

$$M = Nm \tag{7.14}$$

Alternatively we could use (7.14) which expresses the fact that the mass of the Universe is given by the mass of the N elementary particles in it and deduce equations (7.11), (7.12) and (7.13). Indeed a rationale for this is the fact that the Universe at large is electrically neutral and so it is the gravitational force which predominates, and this is very weak in comparison to Electromagnetism. Using the expressions for the Planck length as a Compton wavelength and equating it to (7.4) we can easily deduce

$$Gm^2 = \frac{e^2}{n} = \frac{e^2}{\sqrt{N}} \tag{7.15}$$

wherein we have also used $\hbar c \sim e^2$ and (7.6). Equation (7.15) is as we saw earlier an empirically well known equation. Interestingly, to re-emphasize, as we have deduced (7.15), rather than use it empirically, this points to a unified description of Electromagnetism and Gravitation. We shall explore this relation further.
Interestingly also rewriting (7.13) as

$$G = \frac{l^2c^2}{Rm}$$

wherein we have used (7.1) and further using the fact that $H = c/R$, where H is the Hubble constant we get, as already deduced

$$m \approx \left(\frac{H\hbar^2}{Gc}\right)^{\frac{1}{3}} \tag{7.16}$$

Equation (7.16) is the so called mysterious Weinberg formula, known empirically and encountered earlier [17]. As Weinberg put it, "...it should be noted that the particular combination of $\hbar, H, G$, and c appearing (in the formula) is very much closer to a typical elementary particle mass than other random combinations of these quantities; for instance, from $\hbar, G$, and c alone one can form a single quantity $(\hbar c/G)^{1/2}$ with the dimensions of a mass, but this has the value $1.22 \times 10^{22} MeV/c^2$, more than a typical particle mass by about 20 orders of magnitude!

"In considering the possible interpretations (of the formula), one should be careful to distinguish it from other numerical "coincidences"... In contrast, (the formula) relates a single cosmological parameter, H, to the fundamental constants $\hbar, G, c$ and m, and is so far unexplained."
We remark that (7.13) brings out gravitation in a different light—–somewhat but not exactly, on the lines of Sakharov. In fact it shows up gravitation as the excess or residual energy in the universe. We will return to this in the sequel.
Finally it may be observed that (7.13) can also be rewritten as

$$N = \left(\frac{c^2 l}{mG}\right)^2 \sim 10^{80} \tag{7.17}$$

and so also (7.10) can be rewritten as

$$n = \left(\frac{lc^2}{Gm}\right) \sim 10^{40}$$

It now immediately follows that

$$\bar{N} \sim 10^{120}$$

Looking at it this way, given G and the microphysical parameters we can deduce the large scale numbers $N, \bar{N}$ and n!

7.2 Remarks

The many so called large number coincidences and the mysterious Weinberg formula can be deduced on the basis of a Planck scale underpinning for the elementary particles and the whole Universe. This as we saw in Chapter 3, was done from a completely different point of view, namely using fuzzy spacetime and fluctuations in a 1997 model which as we saw successfully predicted a dark energy driven accelerating universe with a small cosmological constant [175, 130].
However the above treatment brings out the role of the Planck scale particles in the Quantum vaccuum. It resembles to a certain extent, as remarked earlier the Sakharov-Zeldovich metric elasticity of space approach [278]. Essentially Sakharov argues that the renormalization process in Quantum Field Theory which removes the Zero Point energies is altered in General Relativity due to the curvature of spacetime, that is the renormalization or subtraction no longer gives zero but rather there is a residual energy similar to the modification in the molecular bonding energy due to deformation of

the solids. We see this in a little more detail following Wheeler [45]. The contribution to the Lagrangian of the Zero Point energies can be given in a power series as follows

$$L(r) = A\hbar \int k^3 dk + B\hbar^{(4)}r \int kdk$$

$$+\hbar[C(^{(4)}r)^2 + Dr^{\alpha\beta}r_{\alpha\beta}] \int k^{-1}dk$$

$$+(\text{higher-order terms}). \tag{7.18}$$

where A, B, C etc. are of the order of unity and r denotes the curvature. By renormalization the first term in (7.18) is eliminated. According to Sakharov, the second term is the action principle term, with the exception of some multiplicated factors. (The higher terms in (7.18) lead to corrections in Einstein's equations.) Finally Sakharov gets

$$G = \frac{c^3}{16\pi B\hbar \int kdk} \tag{7.19}$$

Sakharov then takes a Planck scale cut off for the divergent integral in the denominator of (7.19). This immediately yields

$$G \approx \frac{c^3 l_P^2}{\hbar} \tag{7.20}$$

In fact using relations like (7.1), (7.6) and (7.12), it is easy to verify that (7.20) gives us back (7.10) (and (7.13)).

According to Sakharov (and (7.20)), the value of G is governed by the Physics of Fields and Particles and is a measure of the metrical elasticity at small spacetime intervals. It is a microphysical constant.

However in our interpretation of (7.13), G appears as the expression of a residual energy over the entire universe: The entire universe has an underpinning of the $\bar{N}$ Planck oscillators and is made up of N elementary particles, which again each have an underpinning of n Planck oscillators. It must be reiterated that (7.20) obtained from Sakharov's analysis shows up G as a microphysical parameter because it is expressed in their terms. This is also the case in Dirac's cosmology. This is also true of (7.10) because n relates to the micro particles exclusively.

However when we use the relation (7.12), which gives n in terms of N, that is links up the microphysical domain to the large scale domain, then we get (7.13). With Sakharov's equation (7.20), the mysterious nature of the Weinberg formula remains. But once we use (7.13), we are effectively using the large scale character of G−−it is not a microphysical parameter. This is brought out by (7.17), which is another form of (7.13). If G were a microphysical parameter, then the number of elementary particles in the universe would depend solely on the microphysical parameters and would not be a large scale parameter. The important point is that G relates to elementary particles and the whole universe [400]. That is why (7.13) or equivalently the Weinberg formula (7.16) relate supposedly microphysical parameters to a cosmological parameter. Once the character of G as brought out by (7.13) is recognized, the mystery disappears. We will also touch upon this point in the next Chapter.

Finally as remarked attempts to unite Gravitation with other interactions have been unsuccessful for several decades. However, it is possible to get a description of Gravitation in an extended gauge field formulation using noncommutative geometry (to take account of the fact that the graviton is a spin 2 particle) [385, 16].

7.3 Gravitation and Black Hole Thermodynamics Again

A few decades ago, the work of Hawking, Beckenstein (and Unruh) and others brought out the connection between Thermodynamics, black holes and Quantum Theory. We will now return to the striking parallel between the disparate fields of Gravitation and thermodynamical considerations on the one hand and Electromagnetism on the other. We will then investigate the mechanism that leads to such a parallelism.

Our starting point is the well known relation (7.15) between the gravitational and electromagnetic coupling constants encountered repeatedly [45]

$$\frac{Gm^2}{e^2} = \frac{1}{\sqrt{N}} \tag{7.21}$$

In (7.21), m is the mass of a typical elementary particle and $N \sim 10^{80}$ is the number of elementary particles in the universe. Equation (7.21) is one of the Dirac large number relations and for this purpose it does not really matter if m stands for the mass of a pion or a proton or an electron (Cf.[45]). It may also be mentioned that (7.21) was considered to be a miraculous large number coincidence along with a few other such relations. However we have already seen in Chapter 3 that these relations can in fact be deduced from the theory [175, 176, 277, 130, 253, 401]. As such they are not empirical or accidental.

In our scheme we deduced that

$$G = \frac{lc^2}{\sqrt{N}m} = \frac{lc^2\tau}{mt} \equiv \Theta/t \tag{7.22}$$

where l is the Compton wavelength of a typical elementary particle $\sim 10^{-13}cms, \Theta \sim 10^9$ and wherein we have used the relation $T = \sqrt{N}\tau$. Equation (7.22) shows the dependence of G on time, and leads to meaningful observational consequences including the otherwise unexplained anomalous accelerations of the Pioneer spacecrafts [208, 335]. We saw all this in Chapter 3. Equation (7.22) is just another form of (7.21). It was also pointed out [277] that (7.22) shows up Gravitation as an effect of Electromagnetism spread over the N particles of the universe. It should be mentioned, as we will see a little later that in (7.22) if we use the fact that $\hbar c/e^2 = 137$ and the Weyl-Eddington relation, we recover (7.21). As we will see (equations (7.25) and following relations of Black Hole Thermodynamics), (7.22) plays an important role.

We would like to stress that if $N \sim 1$ then Gm^2 can be replaced by e^2. This signifies the fact that at the Planck scale, that is for Planck mass black holes all of the electromagnetic energy is of the same order as the gravitational energy or vice versa. Carrying this out on (7.22) we get, as indeed we saw earlier,

$$e^2 = lmc^2 \, \text{or} \, l = e^2/mc^2 \tag{7.23}$$

Apart from the fact that (7.23) is known to be correct, it also follows by a simple substitution of (7.21) in (7.22).

Let us now contrast the gravitational and electromagnetic aspects. For a Planck mass the Schwarzschild radius is the Planck length or Compton length for a Planck mass, as we have seen:

$$\frac{Gm_P}{c^2} = l_P \sim \hbar/m_P c \sim 10^{-33}cm \tag{7.24}$$

We can compare (7.24) with (7.23) which defines l as what we have called the "electromagnetic Schwarzschild" radius viz., the Compton wavelength, when e^2 is seen as an analogue of Gm^2. To push these considerations further, we have from the theory of black hole thermodynamics [44, 402] for any arbitrary mass m, that first the Beckenstein temperature is given by

$$T = \frac{\hbar c^3}{8\pi kmG} \tag{7.25}$$

This was the work of Jacob Beckenstein in the seventies, of the Beckenstein-Hawking radiation fame. Equation (7.25) gives the thermodynamic temperature of a Planck mass black hole. Further, in this theory, as already noted,

$$\frac{dm}{dt} = -\frac{\beta}{m^2}, \tag{7.26}$$

where β is given by

$$\beta = \frac{\hbar c^4}{(30.8)^3 \pi G^2}$$

This leads back to the usual black hole life time given by

$$t = \frac{1}{3\beta} m^3 = 8.4 \times 10^{-24} m^3 secs \tag{7.27}$$

Let us now factor in the time variation of G into (7.26) as we did before. Essentially we use (7.22). Equation (7.26) now becomes

$$m^2 dm = -B\,\mu^{-2} t^2 dt, B \equiv \frac{\hbar c^4}{\lambda^3 \pi}, \mu \equiv \frac{lc^2 \tau}{m}, \lambda^3 = (30.8)^3 \pi$$

Whence on integration we get

$$m = \frac{\hbar}{\lambda \pi^{1/3}} \left\{ \frac{1}{l^6} \right\}^{1/3} t = \frac{\hbar}{\lambda \pi^{1/3}} \frac{1}{l^2} t \tag{7.28}$$

If we use the pion mass, m in (7.28), we get for t, the pion Compton time. In fact if we use (7.22) in (7.25) with the appropriate expression for Θ, we get

$$kT = \frac{mc^3 t}{l}$$

Using for t, the pion Compton time, we get for a typical elementary particle,

$$kT = mc^2 \tag{7.29}$$

We saw that equation (7.29) is the well known relation expressing the Hagedorn temperature of elementary particles [253]. It is an analogue of (7.25). Alternatively if we carry out the substitution $Gm^2 \to e^2$ in (7.25) in the above, we recover (7.29). Similarly instead of (7.26) we will get, with such a substitution,

$$\frac{dm}{dt} = -\frac{\hbar c^4}{\lambda^3 e^4} m^2,$$

whence we get for the life time

$$\frac{\hbar c^4}{\lambda^3 e^4} t = \frac{1}{m} \tag{7.30}$$

Coupling : Gm^2	Coupling : e^2
Schwarzchild radius	Compton wavelength
Beckenstein Temperature	Hagedorn temperature
Beckenstein decay time	Compton time
Planck mass m_P	Pion mass m_π

For an elementary particle, (7.28) and (7.30) are the same. Further from (7.30) we get, for the pion,

$$t \sim 10^{-23} secs,$$

which is again the pion Compton time. So the Compton time shows up as an "electromagnetic Beckenstein radiation life time."

Thus for elementary particles, working within the context of gravitational theory, but with a time varying Gravitational constant being taken into consideration as in steps leading to (7.28), we get the meaningful relations (7.23) and (7.28) and (7.29) giving the Compton length and Compton time as also the Hagedorn temperature as the analogues of the Schwarzschild radius, radiation life time and black hole temperature obtained with the usual gravitational coupling constant. Equivalently we can deduce the same results by scaling up the gravitational coupling constant $Gm^2 \to e^2$. The converse holds good for $e^2 \to Gm^2$. The parallel is complete. The analogy can be summarized as shown in Table 1.

7.4 Further Remarks

1. We note that the relation

$$\frac{\lambda^3 e^4}{\hbar^2 c^2} \sim 1$$

which follows from (7.30) on using the expression for the Compton time for t gives an estimate for the fine structure constant $\sim 1/150$. It must be remarked that earlier, we could see that h itself can be characterized in terms of fluctuations.

2. We have noticed that Gravitation in a sense is a form of weak Electromagnetism. A question that has perplexed us for over a century is, why is Gravitation so much weaker than Electromagnetism– to the extent given by (7.21), in fact. One way in which this can be understood is by realizing that the universe is by and large electrically neutral, because the atoms consist of an equal number of positive and negative charges. Strictly speaking

atoms are therefore electrical dipoles.

With this background and in the light of considerations in Chapter 1, let us consider the following simple model of an electrically neutral atom which nevertheless has a dipole effect. In fact as is well known from elementary electrostatics the potential energy at a distance r due to the dipole is given by

$$\phi = \frac{\mu}{r^2} \tag{7.31}$$

where $\mu = eL, L \sim 10^{-8}cm \sim 10^3 l \equiv \omega l$, e being the electric charge of the electron for simplicity and l being the electron Compton wavelength. (There is a factor $cos\Theta$ with μ, but on an integration over all directions, this becomes an irrelevant constant factor 4π.)

Due to (7.31), the potential energy of a proton p (which approximates an atom in terms of mass) at the distance r (much greater than L) is given by

$$\frac{e^2 L}{r^2} \tag{7.32}$$

As there are $N \sim 10^{80}$ atoms in the universe, the nett potential energy of a proton due to all the dipoles is given by

$$\frac{Ne^2 L}{r^2} \tag{7.33}$$

In (7.33) we use the fact that the predominant effect comes from the distant atoms which are at a distance $\sim r$, the radius of the universe.

We next use the Eddington formula encountered several times earlier,

$$r \sim \sqrt{N} l \tag{7.34}$$

If we introduce (7.34) in (7.33) we get, as the energy E of the proton under consideration

$$E = \frac{\sqrt{N} e^2 \omega}{r} \tag{7.35}$$

Let us now consider the gravitational potential energy E' of the proton p due to all the other N atoms in the universe. This is given by

$$E' = \frac{GMm}{r} \tag{7.36}$$

where m is the proton mass and M is the mass of the universe.

Comparing (7.35) and (7.36), not only is E equal to E', but remembering that $M = Nm$, we get back equation (7.21),

$$\frac{e^2}{Gm^2} = \frac{1}{\sqrt{N}}$$

Thus we have shown that the well known relations and parameters pertaining to Planck mass Black Hole Thermodynamics go over to corresponding elementary particle relations, as shown in Table 1, when the time variation of the gravitational constant as given in (7.22) is invoked. It has also been mentioned that the time variation given in (7.22) has been confirmed by observations on earth, the solar system and, perhaps the most accurate of all, Pulsars (though observations based on other models have yielded slightly different results). We had seen all this in Chapter 3.

7.5 Gravitation From Fluctuations

Richard Feynman had said that gravitation has to be the fluctuation of something [403]. Our earlier work had implied that gravitation can indeed come out as the fluctuation of electromagnetic energy of the N particles comprising the universe (or more correctly, the charges in them). This is given by

$$E = \frac{e^2\sqrt{N}}{R} \tag{7.37}$$

Let us identify E with the gravitational energy of the N particles:

$$E = \frac{Gm^2 N}{R} \tag{7.38}$$

We then get back equation (7.15) viz.,

$$e^2/(Gm^2) = \sqrt{N}$$

We are stressing repeatedly that (7.15) is not an accidental Large Number relation, as it had been supposed, but is an expression of the character of Gravitation itself as being non-fundamental, unlike Electromagnetism. Rather it arises from Electromagnetism of all other particles. Gravitation is now unified with Electromagnetism as expressed by (7.15), but in an unexpected sense.

Incidentally, if we identify (7.38) with the inertial energy of a particle, we get back our earlier relation,

$$G = \frac{lc^2}{\sqrt{N}m},$$

expressing the distributional (or time varying) character of G.

Similarly, (7.37) gives either an expression for the fine structure constant as,

$$e^2/\hbar c = m_e/m_\pi,$$

remembering that strictly speaking, l refers to the Compton length of a pion, whereas m in the fluctuational energy of electric charges refers to the electron mass. Alternatively, we get

$$e^2/l = mc^2$$

which identifies the inertial energy, essentially as the electromagnetic self energy, as discussed in Chapter 1.

A further remark is in order. Earlier, we had included Gravitation in a gauge like formulation, retaining terms of the order of l^2. It must be emphasized that this again is directly linked to large scale fluctuations in a thermodynamic sense, as we saw and, as expressed, for example by the Weyl-Eddington formula.

Chapter 8

An Adventurer's Miscellany

8.1 "Scaled" Quantum Mechanics

Christopher Columbus, the sixteenth century Italian adventurer is a metaphor for how wrong reasons could lead to results which are right, in an unexpected way. His proposal to make a trip to India was rejected by the learned scholars of Salamanaca. Rightly, they pointed out that he had completely underestimated the earth's circumference. Columbus, in fact was using data from the ancient Greeks! There is another story about a royal ball. Several of the important guests were discussing hotly how to make a boiled egg stand upright. "Easy," said Columbus. He bit off the tip and made the egg stand. The spirit of Columbus can open new horizons!
Let us continue in this vein. We will first argue that there is a manifestation of what may be called "scaled" Quantum Mechanics, at different scales in the Universe, and not just at the usual Quantum scale.
We have already argued that in the Universe at large, there appear to be the analogues of the Planck constant at different scales [404, 405]. Infact we have

$$h_1 \sim 10^{93} \tag{8.1}$$

for super clusters;

$$h_2 \sim 10^{74} \tag{8.2}$$

for galaxies and

$$h_3 \sim 10^{54} \tag{8.3}$$

for stars. And

$$h_4 \sim 10^{34} \tag{8.4}$$

for Kuiper Belt objects. In equations (8.1) - (8.4), the $h_\imath$ play the role of the Planck constant, in a sense to be described below. The origin of these equations is related to the following empirical relations

$$R \approx l_1 \sqrt{N_1} \tag{8.5}$$

$$R \approx l_2 \sqrt{N_2} \tag{8.6}$$

$$l_2 \approx l_3 \sqrt{N_3} \tag{8.7}$$

$$R \sim l \sqrt{N} \tag{8.8}$$

and a similar relation for the KBO (Kuiper Belt objects)

$$L \sim l_4 \sqrt{N_4} \tag{8.9}$$

where $N_1 \sim 10^6$ is the number of superclusters in the Universe, $l_1 \sim 10^{25} cms$ is a typical supercluster size, $N_2 \sim 10^{11}$ is the number of galaxies in the Universe and $l_2 \sim 10^{23} cms$ is the typical size of a galaxy, $l_3 \sim 1$ light years is a typical distance between stars and $N_3 \sim 10^{11}$ is the number of stars in a galaxy, R being the radius of the Universe $\sim 10^{28} cms, N \sim 10^{80}$ is the number of elementary particles in the Universe and l is the pion Compton wavelength and $N_4 \sim 10^{10}, l_4 \sim 10^5 cm$, is the dimension of a typical KBO (with mass $10^{19} gm$ and L the width of the Kuiper Belt $\sim 10^{10} cm$ cf.ref.[130]).

The size of the Universe, the size of a supercluster etc. from equations like (8.5)-(8.9), as described in the references turn up as the analogues of the Compton wavelength. For example we have

$$R = \frac{h_1}{Mc} \tag{8.10}$$

where M is the mass of the universe. One can see that equations (8.1) to (8.10) are a consequence of gravitational orbits (or the Virial Theorem) and the conservation of angular momentum viz.,

$$\frac{GM}{L} \sim v^2, MvL = H \tag{8.11}$$

(Cf.refs.[404, 405]), where L, M, v represent typical length (or dispersion in length), mass and velocities at that scale and H denotes the scaled Planck constant.

It also appears that equations (8.5) to (8.9) resemble a typical Random Walk relation (Cf.[108]) of Brownian motion which we encountered in Chapters 2 and 3.

All this is suggestive but empirical. The question arises whether there is any theoretical justification. To investigate this further we observe that if we use (8.11) along with the relation,

$$L = vT$$

where T is a typical time scale, for example the time period for an orbit, we get the relations

$$L^2 = \frac{H}{M}T \quad \left(H = \frac{GM^2}{v}\right) \tag{8.12}$$

(8.12) is the analogue of the well known diffusion equation encountered in Chapter 2 viz.,

$$\Delta x^2 = \nu \Delta t, \quad \nu = \frac{h}{m} \tag{8.13}$$

where ν is the diffusion constant, h the Planck constant and m the mass of a typical particle.

We now observe that as we saw, the relations (8.12) or (8.13) lead to an equation identical to the Quantum Mechanical Schrödinger equation (Cf.ref.[131] for a detailed derivation)

$$h_\imath \frac{\partial \psi}{\partial t} + \frac{h_\imath^2}{2m}\nabla^2 \psi = 0 \tag{8.14}$$

(for different $h_\imath$). Indeed this is not surprising because one can rewrite equation (8.13) as

$$m\Delta x \frac{\Delta x}{\Delta t} = h = \Delta x \cdot \Delta p \tag{8.15}$$

which gives the Uncertainty relation. Conversely, from the Uncertainty Principle (8.15) we could get back (8.12) or (8.13).

Interestingly it has been shown that this is true, not just for the special form of the diffusion constant, but also for any other form of the diffusion constant [406]. Another interesting point is that starting from (8.12) or (8.13), we can deduce equations like (8.5), which describe a Brownian path [157].

In any case the steps leading to equation (8.14) and (8.14) itself provide the rationale for the scaled De Broglie or Compton lengths, for example equation (8.10), which follow from (8.15).

All this can be linked to Critical Point Theory and the Renormalization Group exactly as earlier. Relations like (8.5) to (8.9) would then be the result of equations like

$$\bar{Q}^\nu = \xi^\beta \,\text{and}\, \bar{Q} \sim \frac{1}{\sqrt{N}}, \xi = (l/R)^2 \tag{8.16}$$

which we encountered earlier at different scales.
We also observe that a Schrödinger equation like procedure has been used though in an empirical way by Agnese and Festa [407] to derive a Titius-Bode type relation for planetary distances which now appear as quantized levels. This consideration has been extended in an empirical way to also account for quantized cosmic distances [408].
Interestingly if we consider a wave packet of the generalized Schrödinger equation (8.14) with h_1 given by (8.1) for the Universe itself, we have for a Gaussian wave packet

$$R \approx \frac{\sigma}{\sqrt{2}} \left(1 + \frac{h_1^2 T^2}{\sigma^4 M^2}\right)^{1/2} \left(\approx \frac{1}{\sqrt{2}} \frac{h_1 T}{\sigma M}\right)$$

where R and T denote the radius and age of the Universe, M its mass and $\sigma \sim R$ is the spread of the wave packet in the spirit of Chapter 2. As $R \approx cT$ this gives us back (8.10), that is the "Compton wavelength" of the Universe treated as a wave packet.
Interestingly also we can pursue the reasoning of equations like (8.1) to the case of terrestrial phenomena. Let us consider a gas at standard temperature and pressure. In this case, the number of molecules $n \sim 10^{23}$ per cubic centimeter, so that $r \sim 1cm$ and with the same l, we can get a "scaled" Planck constant $\tilde{h} \sim 10^{-44} << h$, the Planck constant. That is, we come back to the classical case.
In this case, a simple application of the WKB approximation, leads immediately from the Schrödinger equation at the new scale to the classical Hamilton-Jacobi theory, that is to classical mechanics.
Equations like (8.5) are the analogue of the well known Eddington formula. Similarly we can have the analogue of the mysterious Weinberg relation linking the pion mass to the Hubble constant, from $H^2 = M^3 LG$. For this we need to define the analogue of the Hubble constant H

$$\hat{H} = \frac{v}{L}$$

to get

$$M = \left(\frac{\hat{H} H^2}{Gv}\right)^{\frac{1}{3}}$$

which is the required relation.
We can now argue that just as matter in the form of elementary particles, forms or condenses within the Compton wavelength from a background Quantum vacuum in a phase transition, matter at other scales, for example

stars and galaxies also could be considered to condense or cluster by a similar mechanism. This would give a rationale for the observed lumpiness of the Universe. Similar considerations apply for the other scales referred to.

We will now argue afresh, following the above reasoning, that the difference between Electromagnetism at the micro scale and Gravitation at the macro scale is merely a matter of the difference in the time and length scales. While we saw what follows earlier, the context now is that an underlying principle operates at different scales in the Universe.

Infact the operative equations are (8.16):

$$\bar{Q}^{\nu} = \bar{\xi}^{\beta}$$

where $\bar{Q}$ and $\bar{\xi}$ are the reduced order parameter and correlation length. We now have

$$\bar{Q} \sim \frac{1}{\sqrt{N}}, \bar{\xi} = (l/R)^2$$

which gives, the Eddington like relations.

Now if we consider the representation of the Hamiltonian as the differential time operator we will get

$$H(T) = \frac{d}{dT} = \frac{d}{\sqrt{N} d\tau} = \frac{H(\tau)}{\sqrt{N}} \tag{8.17}$$

$H(T)$ in (8.17) denotes Gravitation represented by the coupling constant Gm^2 and $H(\tau)$ in (8.17) denotes Electromagnetism represented by the coupling constant e^2 and m referring to the same elementary particle. The rationale for (8.17) is that Gravitation operates at the scale of the universe, whereas Electromagnetism operates at the elementary particle scale. Whence if (8.17) is consistent, we should have,

$$\frac{e^2}{Gm^2} \sim \sqrt{N} \tag{8.18}$$

In fact this is the well known empirical and supposedly accidental relation which we saw several times—–the ratio of the coupling constants encountered earlier.

Let us now consider the analogue of the microscopic relation,

$$m \frac{l^2}{\tau} = h$$

for the macro or cosmic scale. We then get

$$h \to ML^2/T = h_1 \sim 10^{93} \tag{8.19}$$

This equation which is the same as (8.1), is in fact perfectly meaningful because h_1 in (8.19) is the Godel spin of the Universe [404, 408]. In fact (8.19) immediately leads to

$$R = \frac{h_1}{Mc} \tag{8.20}$$

which is (8.10). Equations (8.19) and (8.20) again show that the Universe itself seems to follow a Quantum Mechanical behaviour with a scaled up Planck constant h_1 as argued previously.

The above considerations in the context of universality and scaling effects of Critical Point Phenomena and the Renormalization Group mean: The Universe is a coarse grained scaled up version of the micro world, and Gravitation being the counterpart of Electromagnetism should be given by their mutual scaled ratio. Let us see if this model is correct.

In such a coarse graining, we know that at a Critical Point we have for the coupling constants,

$$J^{(1)}/kT_c^{(1)} = 1 \; J^{(2)}/kT_c^{(2)} = 1$$

where from the theory, in our case,

$$T_c^{(1)}/T_c^{(2)} = l/R$$

Whence we get

$$J^{(1)}/J^{(2)} = l/R \tag{8.21}$$

As $J^{(1)} = Gm^2$ and $J^{(2)} = e^2$ are the coupling constants at the two scales, does (8.21) give the correct ratio? In fact it gives us back (8.18). In other words, as can be seen from (8.17) or (8.21), the "weak" gravitational interaction is a manifestation of the much longer time periods involved on the macro or cosmic scale, while the much stronger electromagnetic interaction is a manifestation of the much smaller scale of time at the micro level. This can be elaborated upon in the following way.

The electromagnetic energy of a typical elementary particle, for example the pion is given by

$$\text{Energy} = \frac{e^2}{l} = \frac{\hbar}{\tau}$$

On the other hand its gravitational energy is given by

$$\text{Gravitational Energy} = \frac{Gm^2}{l} = \frac{\hbar}{T} \tag{8.22}$$

Whence,

$$\frac{Gm^2}{e^2} = \frac{l}{R} = \frac{1}{\sqrt{N}} \tag{8.23}$$

which is again, (8.18). In both these cases, as we have been dealing with a microscopic particle, the Heisenberg Uncertainty Principle holds. So while the electromagnetic energy plays out in the Compton time τ, the gravitational energy plays out during the life time of the Universe. Sivaram [140] uses in (8.22) the relation $T = \frac{1}{H}$, where H is the Hubble constant, to get, as a curiosity, the mysterious Weinberg formula again. We will return to the "scaling" in (8.23) later.

Let us now consider the gravitational energy of all the N particles in the Universe. This is given by

$$E = \frac{NGm^2}{l}$$

The energy E has a low Beckenstein temperature and as can be easily calculated from the Beckenstein Radiation decay formula viz.,

$$T = 8.4 \times 10^{-24} (E/c^2)^3$$

the life time is T, the age of the Universe itself.

Interestingly if the above considerations are carried over to the Planck scale versus the Compton scale, we can easily verify that there is no new scaled down Planck constant, as for example in (8.19)−−that is the considerations remain the same as those at the Compton scale. However, let us see what we get if in analogy to (8.16) and (8.21) we compare the Planck and Compton scales. This time, the Critical Point relations lead to the known relation,

$$l = \sqrt{n} l_P, \quad \tau = \sqrt{n} \tau_P.$$

Furthermore, (8.17), with a similar notation leads to,

$$H(\tau) = \frac{H(\tau_P)}{\sqrt{n}}$$

which also we have encountered earlier. It is just,

$$m = m_P / \sqrt{n}$$

Further, the Beckenstein Radiation life time of a Planck mass, gives this time−−the Planck Compton time. We can look at the above in another way. We have already seen that an elementary particle is the result of a fluctuation of n Planck oscillators while the Universe is made up of the fluctuation of $\bar{N} \sim 10^{120}$ such oscillators. So the elementary particle electromagnetic interaction bears to the "universal" gravitational interaction, a ratio,

$$H_{emag}/H_{grav} = \sqrt{\bar{N}}/\sqrt{n},$$

which is the same as (8.18).
This can be illustrated by the following amusing description in Indian Mythology. Brahma, the creator of the Universe has a very very long day——while he takes a bath, many time consuming and momentous events take place on the earth. By Brahma's reckoning, however, the time elapsed is still miniscule. Interestingly the ratio of the time scales would be the same as above, because of the fact that the estimate for the age of the Universe or Brahma's day is exactly of the same order of magnitude as modern estimates.
We may also add that as seen earlier, the neutrino may be thought consistently to have a scaled Planck constant $10^{-12}\hbar$.

8.2 Quantum Geometry I

One of the earliest attempts to unify electromagnetism and gravitation, was, as we saw, Weyl's gauge invariant geometry. The basic idea was [38] that while

$$ds^2 = g_{\mu\nu}dx^\mu dx^\nu \tag{8.24}$$

was invariant under arbitrary transformations in General Relativity, a further invariant, namely,

$$\Phi_\mu dx^\mu \tag{8.25}$$

which is a linear form should be introduced. $g_{\mu\nu}$ in (8.24) would represent the gravitational potential, and Φ_μ of (8.25) would represent the electromagnetic field potential. As Weyl observed, "The world is a $3+1$ dimensional metrical manifold; all physical field–phenomena are expressions of the metrics of the world. (Whereas the old view was that the four-dimensional metrical continuum is the scene of physical phenomena; the physical essentialities themselves are, however, things that exist "in" this world, and we must accept them in type and number in the form in which experience gives us cognition of them: nothing further is to be "comprehended" of them.)$\cdots$"
This was a bold step, because it implied the relativity of magnitude multiplied effectively on all components of the metric tensor $g_{\mu\nu}$ by an arbitrary

function of the coordinates. However, the unification was illusive because the $g_{\mu\nu}$ and Φ_μ were really independent elements. As Einstein noted, in Stafford Little Lectures delivered in May 1921 at Princeton University [378], "...if we introduce the energy tensor of the electromagnetic field into the right hand side of (the gravitational field equation) we obtain (the first of Maxwell's systems of equations in tensor density form), for the special case $(\sqrt{-g}\rho\frac{dx_\nu}{ds} =)r^\mu = 0, \cdots$ This inclusion of the theory of electricity in the scheme of General Relativity has been considered arbitrary and unsatisfactory... a theory in which the gravitational field and the electromagnetic field do not enter as logically distinct structures would be much preferable..." A more modern treatment is recapitulated below [39].
The above arbitrary multiplying factor is normalized and we require that,

$$|g_{\mu\nu}| = -1, \tag{8.26}$$

For the invariance of (8.26), $g_{\mu\nu}$ transforms now as a tensor density of weight minus half, rather than as a tensor in the usual theory. The covariant derivative now needs to be redefined as

$$T^{\iota\cdots}_{\kappa\cdots\,;\sigma} = T^{\iota\cdots}_{\kappa\cdots\,,\sigma} + \Gamma^\iota_{\rho\sigma}T^{\rho\cdots}_{\kappa\cdots} - \Gamma^\rho_{\kappa\sigma}T^{\iota\cdots}_{\rho\cdots} - nT^{\iota\cdots}_{\kappa\cdots}\Phi_\sigma, \tag{8.27}$$

In (8.27) we have introduced the Φ_μ, and n is the weight of the tensor density. This finally leads to (Cf.ref.[39] for details).

$$\Phi_\sigma = \Gamma^\rho_{\rho\sigma}, \tag{8.28}$$

Φ_μ in (8.28) is identified with the electromagnetic potential, while $g_{\mu\nu}$ gives the gravitational potential as in the usual theory. We had encountered (8.28) earlier via a different route. The affine connection is now given by

$$\Gamma^\lambda_{\iota\kappa} = \frac{1}{2}g^{\lambda\sigma}(g_{\iota\sigma,\kappa} + g_{\kappa\sigma,\iota} - g_{\iota\kappa,\sigma}) + \frac{1}{4}g^{\lambda\sigma}(g_{\iota\sigma}\Phi_\kappa + g_{\kappa\sigma}\Phi_\iota - g_{\iota\kappa}\Phi_\sigma) \equiv \begin{pmatrix} \lambda \\ \iota\kappa \end{pmatrix} \tag{8.29}$$

The essential point, and this was the original criticism of Einstein and others, is that in (8.29), $g_{\mu\nu}$ and Φ_μ are independent entities.
Let us now analyze the above from a different perspective. Let us again write the product $dx^\mu dx^\nu$ of (8.24) as a sum of half its anti-symmetric part and half the symmetric part. The invariant line element in (8.24) now becomes $(h_{\mu\nu} + \hbar_{\mu\nu})dx^\mu dx^\nu$ where h and $\hbar$ denote the anti-symmetric and symmetric parts respectively of g. h would vanish unless the commutator

$$[dx^\mu, dx^\nu] \approx l^2 \neq 0 \tag{8.30}$$

l being some fundamental minimum length. In fact h can be characterized as

$$h_{\mu\nu} = \eta^{\rho\sigma}\epsilon_{\rho\sigma\mu\nu},$$

where η is an antisymmetric tensor and ϵ is the Levi-Civita tensor density. As pointed out earlier the noncommutative geometry given in (8.30) was studied by Snyder and others though from a different perspective. We have already shown in Chapter 6, in detail that under a time elapse transformation of the wave function, (or, alternatively, as a small scale transformation), we recover the Dirac equation, given the geometry (8.30) (Cf. also [107, 371]).
The Dirac wave function itself is given by

$$\psi = \begin{pmatrix} \chi \\ \Theta \end{pmatrix},$$

where χ and Θ are spinors. We have seen that under reflection while the so called positive energy spinor Θ behaves normally, $\chi \to -\chi, \chi$ being the so called negative energy spinor which comes into play at the Compton scale [181]. That is, as already noted, the space is doubly connected. Because of this property as shown in detail [371], there is now a covariant derivative given by, in units, $\hbar = c = 1$,

$$\frac{\partial\chi}{\partial x^\mu} \to [\frac{\partial}{\partial x^\mu} - nA^\mu]\chi \tag{8.31}$$

where

$$A^\mu = \Gamma^{\mu\sigma}_\sigma = \frac{\partial}{\partial x^\mu} log(\sqrt{|g|}) \tag{8.32}$$

Γ denoting the Christofell symbols.
A^μ in (8.32)is now identified with the electromagnetic potential, exactly as in Weyl's theory except that now, A^μ arises from the bi spinorial character of the Dirac wave function or the double connectivity of spacetime. Further, as we have noted already [129], the mass density of the particle is given by,

$$\rho = \chi\chi^*$$

Indeed ρ vanishes outside the Compton scale for any particle.
What all this means is that the so called ad hoc feature in Weyl's unification theory is really symptomatic of the underlying noncommutative spacetime geometry (8.30). Given (8.30) we get both Gravitation and Electromagnetism in a unified picture, because both are now the consequence of spacetime geometry. We could think that Gravitation arises from the symmetric

part of the metric tensor (which indeed is the only term if $0(l^2)$ is neglected) and Electromagnetism from the antisymmetric part (which manifests itself as an $0(l^2)$ effect). It is also to be stressed that in this formulation, we are working with noncommutative effects at the Compton scale, this being true for the Weyl like formulation also.
We reiterate, once we abandon smooth spacetime manifolds and consider noncommutative geometries defined by, for example (8.30), then we are lead to multiply connected manifolds which conceal the Quantum Mechanical spin half and a unified description of Quantum Mechanics and Geometrodynamics becomes possible. Finally it may be mentioned that the fact that n in (8.31) is integral, explains the discreteness or quantized nature of electric charge.

8.3 Quantum Geometry II

Let us now consider the above ideas in the context of the De Broglie-Bohm formulation [337]. We start with the Schrödinger equation

$$\imath\hbar\frac{\partial\psi}{\partial t} = -\frac{\hbar^2}{2m}\nabla^2\psi + V\psi \tag{8.33}$$

In (8.33), the substitution

$$\psi = Re^{\imath S/\hbar} \tag{8.34}$$

where R and S are real functions of $\vec{r}$ and t, leads as we saw earlier, to,

$$\frac{\partial\rho}{\partial t} + \vec{\nabla}\cdot(\rho\vec{v}) = 0 \tag{8.35}$$

$$\frac{1}{\hbar}\frac{\partial S}{\partial t} + \frac{1}{2m}(\vec{\nabla}S)^2 + \frac{V}{\hbar^2} - \frac{1}{2m}\frac{\nabla^2 R}{R} = 0 \tag{8.36}$$

where

$$\rho = R^2, \vec{v} = \frac{\hbar}{m}\vec{\nabla}S$$

and

$$Q \equiv -\frac{\hbar^2}{2m}(\nabla^2 R/R) \tag{8.37}$$

Using the theory of fluid flow, we also saw that (8.35) and (8.36) lead to the Bohm alternative formulation of Quantum Mechanics (Cf.refs.[409, 323, 410] for a simple treatment). In this theory there is

a hidden variable namely the definite value of position while the so called Bohm potential Q in (8.37) can be non local, two features which do not find favour with physicists. (In our formulation however, the definite value of the position coordinate is fudged by the fuzzyness of spacetime.)
It must be noted that in Weyl's geometry, even in a Euclidean space there is a covariant derivative and a non vanishing curvature R.
Santamato (Cf.refs. [268–271]) exploits this latter fact, within the context of the De Broglie-Bohm theory and postulates a Lagrangian given by

$$L(q,\dot{q},t) = L_c(q,\dot{q},t) + \gamma(\hbar^2/m)R(q,t),$$

He then goes on to obtain the equations of motion like (8.33),(8.34), etc. by invoking an Averaged Least Action Principle

$$I(t_0,t_1) = E\left\{\int_{t_0}^{t} L^*(q(t,\omega),\dot{q}(t,\omega),t)dt\right\}$$
$$= \text{minimum} \qquad (8.38)$$

with respect to the class of all Weyl geometries of space with fixed metric tensor. Equation (8.38) now leads to the Hamilton-Jacobi equation

$$\partial_t S + H_c(q,\nabla S,t) - \gamma(\hbar^2/m)R = 0, \qquad (8.39)$$

Equation (8.39) leads to the Schrödinger equation (in curvilinear coordinates)

$$\imath\hbar\partial_1\psi = (1/2m)\left\{\left[(\imath\hbar/\sqrt{g})\,\partial_\imath\sqrt{g}A_\imath\right]g^{\imath k}\left(\imath\hbar\partial_k + A_k\right)\right\}\psi$$
$$+\left[V - \gamma\left(\hbar^2/m\right)\dot{R}\right]\psi = 0, \qquad (8.40)$$

As can be seen from (8.40), the Quantum potential Q is now given in terms of the scalar curvature R.
We have already related the arbitrary functions Φ of Weyl's formulation with a noncommutative spacetime geometry.
This throws further light on Santamato's postulative approach of extending the De Broglie-Bohm formulation.
At an even more fundamental level, our formalism gives us the rationale for the De Broglie wave length itself. Because of the noncommutative geometry in (8.30) space becomes multiply connected, as we saw, in the sense that a closed circuit cannot be shrunk to a point within the interval. Let us consider the simplest case of double connectivity. In this case, if the interval is of length L, we will have,

$$\Gamma = \int_c m\vec{V}\cdot d\vec{r} = h\int_c \vec{\nabla}S\cdot d\vec{r} = h\oint dS = mV\pi L = \pi h \qquad (8.41)$$

whence

$$L = \frac{h}{mV} \tag{8.42}$$

We had encountered equations like (8.41) earlier. In (8.41), the circuit integral was over a circle of diameter L. Equation (8.42) shows the emergence of the De Broglie wavelength. This follows from the noncommutative geometry of spacetime, rather than the physical Heisenberg Uncertainty Principle. Remembering that Γ in (8.41) stands for the angular momentum, this is also the origin of the Wilson-Sommerfeld quantization rule, an otherwise mysterious Quantum Mechanical prescription. Indeed we have seen all this before.

What we have done is to develop a Quantum Geometrical picture based on the geometry (8.30) and multiply connected spaces.

We finally remark that as seen earlier, the double connectivity of space gives the Quantum Mechanical spin, while the non integrability of the phase gives the electromagnetic field of the particle (Cf. also [298]). Lastly the energy within this region with radius given by the Compton wavelength, viz.

$$\int \rho c^2 d\Omega = mc^2,$$

that is we get the mass, as well (Cf. also ref.[130]).

In other words, as seen in Chapter 6, the considerations of fuzzy spacetime yield at the Compton scale , the mass, spin and electromagnetic field of the elementary particle.

8.4 Large Scale Structures

Our view of the universe has been continuously evolving over the centuries. Thus Newton's universe was one in which the stars were the building blocks. These building blocks were stationary in the universe. After about two centuries this view underwent a transformation, with the discovery in the early twentieth century by Hubble that the so called galactic nebulae were star systems or galaxies, each containing something like a hundred thousand million stars, and these galaxies themselves being at distances far far beyond those of stars. The building blocks were now the galaxies. Then the Red Shift studies of galaxies by V.M. Slipher showed that the galaxies were all rushing outwards. Thus was born the precursor of what has come to be known as the standard Big Bang Cosmology [45]. Soon it was realized that there were clusters of galaxies which would more correctly qualify as the building blocks, and then super clusters. Within clusters and super clusters, there could be departures from the Hubble velocity distance law. The law therefore represented something happening at a very large scale. We got a glimpse of this in Chapter 1.

A further development that came about in the 1980s threw up a dramatically different scenario. The very large clusters of galaxies seemed to lie on bubble or balloon like sheets, there being voids or very thinly populated regions in the interiors. These voids would have dimensions of the order of a hundred million light years [411–414]. This has been a puzzle thrown up in the late twentieth century: Exactly why do we have the voids and why do we have polymer like two dimensional structures on the surfaces of these voids? The puzzle is compounded by the fact that given the dispersion velocities of the galaxies of the order of a thousand kilometers per second, it would still take periods of time greater than the age of the universe, of the order of 13 billion years, for them to move out of an otherwise uniform distribution, leaving voids in their wake. An interesting suggestion was that the galaxies consisting of ordinary matter were floating on the "voids" which are actually made up of dark matter. In any case, latest developments have marginalized dark matter in favour of dark energy.

One of the few explanations for this large scale structure of the universe has been the pancake model of Zeldovich [170]. Essentially according to this model, much of the matter of the universe was in the form of a thin pancake which broke up into pieces, the pieces then forming the clusters of galaxies and galaxies, which have inherited the two dimensional character. Indeed as we briefly noted earlier, studies have suggested this two dimensional character [415].

In the above context we consider the model of "Scaled" Quantum Effects of Section 8.1 [404, 405, 251, 130, 408, 407]. To sum up the main results: It is argued that the structures of the unvierse at different scales mimic Brownian effects, which again lead to a Quantum behavior with different "Scaled" Planck constants. All this is contained in equations (8.1) to (8.10) and subsequent relations.

These considerations lead via the diffusion process to the Schrodinger like equation [131, 16],

$$h_\imath \frac{\partial \psi}{\partial t} + \frac{h_\imath^2}{2m} \nabla^2 \psi = 0 \tag{8.43}$$

for different $h_\imath$ given by equations (8.1) to (8.4).

Before proceeding further we may point out that (8.5) to (8.9) already indicate the two dimensionality referred to above. In fact alternatively, the theory is modelled on a phase transition viz., the Landau-Ginzburg theory applied to an equation like (8.43). Such a phase transition would also explain what the movement of galaxies under normal circumstances cannot, that is the large size of the voids. Under such phase transitions we have equations like (8.16).

Whence we recover the Weyl-Eddington like equations (8.5) to (8.9). This is yet another derivation.

Let us see if the above model can give an estimate for the size of the voids. In fact we have to revert to the Schrodinger like equation (8.43) with a hydrogen like atom, except that GM^2 replaces Ze^2 of the hydrogen atom. Let us consider the hydrogen like wave functions at a scale of galaxies with h_1 replaced by h_2 given above in equation (8.2). The radial part of a typical wave function would be given by [416]

$$\psi_l = \left\{ \left(6 - 6\rho + \rho^2\right) e^{-\rho^{\frac{1}{2}}} \right\} \tag{8.44}$$

It is easy to verify that the expression (8.44) is a maximum with ρ given by,

$$\rho = \frac{2Ze^2 r}{h_2^2} M \to \frac{2GM^3 r}{h_2^2} \sim 10 \tag{8.45}$$

In fact (8.45) gives us back, the Weinberg like formula encountered earlier. We thus have from (8.45) after a simple calculation and feeding in the values for h_2 and $M \sim 10^{44} gm$, that $r \sim 100$ light years, exactly as required. The point is that at radial distances like r given above, there would be a greater concentration of galaxies while within this value of r the distribution would be comparatively sparse. We must also remember that there is the angular part of the "wave function", which means that each value of r really corresponds to a spherical shell.

Thus it is a consequence of Scaled Quantum Effects arising due to the gravitational forces that lead to the bubble and void structure of the universe. Similar arguments could be put forth for the pancake structure of galaxies themselves.

8.5 The Puzzle of Gravitation

We will now argue that the Weinberg formula represents a Machian or holistic effect. Indeed this is suggested by the appearance of the Hubble constant in the expression for the mass of an elementary particle.

Let us consider the gravitational self interaction of a particle (Cf. also ref.[219]). Our starting point is the action functional

$$S = -(8\pi G)^{-1}\int d^4x\phi\Delta^2\phi + \int d^4x\Psi^*\left(\imath\hbar\frac{\partial\Psi}{\partial t} + \frac{\hbar^2}{2m}\Delta^2\Psi - m\phi\Psi\right)$$

where ϕ is some potential whose nature is not as yet specified, G being some coupling constant. The extremum conditions of action with respect to Ψ^* and Ψ lead to the Schrodinger equation with the interaction potential ϕ:

$$\imath\hbar\frac{\partial\Psi}{\partial t} = -\frac{\hbar^2}{2m}\Delta^2\Psi + m\phi\Psi \tag{8.46}$$

and to the Poisson equation for the potential itself

$$\Delta^2\phi = 4\pi Gm\Psi^*\Psi \tag{8.47}$$

Thus, the equations (8.46) and (8.47) describe a self-interacting particle. It is well known that an exact solution to (8.47) is given by

$$\phi(\vec{r},t) = -G\int_\Omega d\Omega(\vec{r})\frac{\rho(\vec{r},t)}{|\vec{r}-\vec{r'}|}, \tag{8.48}$$

where Ω is the three dimensional region which confines the particle, and we have defined

$$\rho(\vec{r},t) = m\Psi^*(\vec{r},t)\Psi(\vec{r},t) \tag{8.49}$$

From (8.48), we can immediately see that for distances far outside the region Ω, that is $|\vec{r}| << |\vec{r'}|$, the potential ϕ has the form

$$\phi \approx \frac{GM}{r}, \tag{8.50}$$

where $r = |\vec{r}|$, and we have defined M as,

$$M = \int_\Omega d\Omega(\vec{r})\rho(\vec{r},t) = m\int_\Omega d\Omega(\vec{r})\Psi^*(\vec{r},t)\Psi(\vec{r},t) \tag{8.51}$$

The integral on the right hand side of (8.51) is conserved in time due to (8.46):

$$\frac{\partial}{\partial t}\int_\Omega d\Omega(\vec{r})\Psi^*(\vec{r},t)\Psi(\vec{r},t) = 0$$

Thus the quantity M is constant, and we can interpret (8.50) and (8.51) as follows. The attractive potential (8.50) is now the classical gravitational potential, M is the gravitational mass, G being the gravitational constant. If we prescribe the unit value to the above conserved functional and interpret it as the norm square, I^2, or the full probability

$$I^2 = \int_\Omega d\Omega(\vec{r})\Psi^*(\vec{r},t)\Psi(\vec{r},t) = 1,$$

then the gravitational mass coincides with the inertial mass,

$$m = M, \tag{8.52}$$

and the quantity (8.49) now can be interpreted as the mass probability density. The source term on the right side of (8.47) is equal to the particle probability density itself.

Now, let us consider the self-consistent problem−−the particle in its own potential well. We cannot obtain an exact solution. However, we can approximately describe some features of such a solution. The first assumption will be that we deal only with a spherically symmetric wave function: $\Psi = \Psi(r,t)$ where r is a radial coordinate. Then the mass probability density has the same dependence: $\rho = \rho(r,t)$. It can be easily shown that for any spherical mass distribution, the potential (8.48) is reducible to a simple form

$$\phi(r,t) = G \int_0^r dr' \frac{m(r',t)}{r'^2} - \int_0^\infty dr' \frac{m(r',t)}{r'^2}, \tag{8.53}$$

where we denote

$$m(r,t) = 4\pi \int_0^r dr' r'^2 \rho(r',t),$$

and $m(r,t)$ is just the mass inside a ball of radius r. Certainly, the solution (8.53) gives an exact formula (8.50) with the mass (8.52) for the point mass distribution. Further, we shall use the value Φ instead of the potential ϕ:

$$\phi(r,t) = mG\Phi(r,t)$$

This allows us to rewrite (8.46) in the form

$$\imath \frac{2m}{\hbar} \frac{\partial}{\partial t} \Psi + \Delta^2 \Psi - \frac{2m^3 G}{\hbar^2} \Phi \Psi = 0 \tag{8.54}$$

The coefficient of Φ in (8.54) has the dimensionality of inverse length. Thus, we denote

$$l_G = \frac{\hbar^2}{2m^3 G}, \tag{8.55}$$

Equation (8.55) is nothing but the Weinberg formula again if we identify l_G with the radius of the Universe and remember that,

$$H = \frac{c}{l_G}$$

All this shows that the mass m of an elementary particle is very Machian, rather than being microphysical, if G is microphysical and vice versa. We, on the other hand, have argued for G being distributioinal, and not microphysical.

8.6 A New Short Range Force

As we saw in Chapter 5, in some ways the General Relativistic gravitational field resembles the electromagnetic field, particularly in certain approximations, as for example when the field is stationary or nearly so and the velocities are small. In this case the equations of General Relativity can be put into a form resembling those of Maxwell's Theory, and then the fields have been called Gravitoelectric and Gravitomagnetic [417]. Experiments have also been suggested for measuring the Gravitomagnetic force components for the earth [418].

We can ask whether such a consideration can be applied to elementary particles, if in fact they can be considered in the context of General Relativity. Apart from Quantum Gravity, there have been different approaches for studying elementary particles via General Relativity [288, 130, 91] and references therein. We will now show that it is possible to extend the Gravitomagnetic and Gravitoelectric formulations to elementary particles within the framework of our theory.

We saw that the linearized General Relativistic equations could describe the properties of elementary particles, such as spin, mass, charge and even the very Quantum Mechanical anomalous gyromagnetic ratio $g = 2$, apart from several other characteristics [299, 324, 175, 312].

We recall that the linearized equations of General Relativity, viz.,

$$g_{\mu\nu} = \eta_{\mu\nu} + h_{\mu\nu}, h_{\mu\nu} = \int \frac{4T_{\mu\nu}(t - |\vec{x} - \vec{x}'|, \vec{x}')}{|\vec{x} - \vec{x}'|} d^3x' \tag{8.56}$$

where as usual,

$$T^{\mu\nu} = \rho u^u u^v \tag{8.57}$$

lead, on using (8.57) in (8.56), to the mass, spin, gravitational potential and charge of an electron, if we work at the Compton scale. Let us now apply the macro Gravitoelectric and Gravitomagnetic equations to the above case. Infact these equations are (Cf.ref.[417]).

$$\nabla \cdot \vec{E}_g \approx -4\pi\rho, \nabla \times \vec{E}_g \approx -\partial\vec{H}_g/\partial t, etc. \tag{8.58}$$

$$\vec{E}_g = -\nabla\phi - \partial\vec{A}/\partial t, \quad \vec{H}_g = \nabla \times \vec{A} \tag{8.59}$$

$$\phi \approx -\frac{1}{2}(g_{00} + 1), \vec{A}_\imath \approx g_{0\imath}, \tag{8.60}$$

The subscripts g in the equations (8.58) and (8.59) are to indicate that the fields E and H in the macro case do not really represent the electromagnetic field, but rather resemble them. Let us apply equation (8.59) to equation (8.56), keeping in mind equation (8.60). We then get, considering only the order of magnitude, which is what interests us here, after some manipulation

$$|\vec{H}| \approx \int \frac{\rho V}{r^2} \bar{r} \approx \frac{mV}{r^2} \tag{8.61}$$

and

$$|\vec{E}| = \frac{mV^2}{r^2} \tag{8.62}$$

V being the speed.

In (8.61) and (8.62) the distance r is much greater than a typical Compton wavelength, to make the approximations considered in deriving the Gravitomagnetic and Gravitoelectric equations meaningful.

Remembering that we have, by the Uncertainty Principle,

$$mVr \approx h,$$

the electric and magnetic fields in (8.61) and (8.62) now become

$$|\vec{H}| \sim \frac{h}{r^3}, |\vec{E}| \sim \frac{hV}{r^3} \tag{8.63}$$

We now observe that (8.63) does not really contain the mass of the elementary particle. Could we get a further insight into this new force?

Indeed in the above linearized General Relativistic characterization of the electron, it turns out as indicated that the electron can be represented by the Kerr-Newman metric which incidentally also gives the anomalous gyromagnetic ratio $g = 2$ of ref. [130]. (This result has recently been reconfirmed by Nottale [419] from a totally different point of view, using scaled relativity.) It is well known that the Kerr-Newman field has extra electric and magnetic terms (Cf.[160]), both of the order $\frac{1}{r^3}$, exactly as indicated in (8.63).

It may be asked if there is any candidate as yet for the above mass independent, spin dependent (through h) short range force. There is already one such experimental candidate—–the inexplicable $B_{(3)}$ [420] short range force, first detected in 1992 at Cornell and since, it is claimed, confirmed by subsequent experiments. It differs from the usual $B_{(1)}$ and $B_{(2)}$ long range fields of Special Relativity.

Interestingly, if we think of the above force as being mediated by a "massive" particle, that is, work with a massive vector field we can recover (8.62) and (8.63) [165]. In this case there is an upper limit on the mass of the photon $\sim 10^{-48}g$, pleasingly, in consonance with our model.

A final comment: It is quite remarkable that equations like (8.58), (8.59) and (8.60) which resemble the equations of Electromagnetism, have in the usual macro considerations no connection whatsoever with Electromagnetism except in appearance. This would seem to be a rather miraculous coincidence. In fact the earlier considerations of linearized General Relativistic theory of the electron as also the Kerr-Newman metric formulation, demonstrate that the resemblance to Electromagnetism is not an accident, because in this latter formulation, both Electromagnetism and Gravitation arise from the metric as noted in Chapter 5 (Cf.also refs.[329, 299, 324, 130]).

8.7 Gravitational Effects

We may next point out the following. Let us introduce the minimum cut off l into the Schwarzchild metric. This gives

$$d\tau^2 = d\tau_0^2 - \frac{2MG}{r}(\frac{l}{r})(dt^2 - dr^2)$$

where $d\tau_0^2$ is the unmodified metric. The above shows that G is replaced by

$$G(1 + \frac{l}{r}).$$

Apart from the fact that this is equivalent to an extra force,

$$\text{Force} \quad \propto \frac{GMl}{r^3},$$

it is also equivalent to the time varying G encountered earlier in Chapter 3 and given by

$$\dot{G} = -G/t$$

The above follows because

$$\frac{r}{t} = \frac{l}{\tau}$$

where r and t are the radius and age of the Universe and l and τ are a typical Compton length and time.

It is interesting to note that in the above analysis, if we take l to be the radius of the Universe and M to be its mass, then the extra force gives the observed cosmological constant. (Interestingly, the Universe itself shows up as a Schwarzchild Black Hole, as noted earlier [130].)

8.8 Bosons as Bound States of Fermions: The Neutrino Universe

In our formulation, Fermions are primary– –Bosons are bound states of Fermions. This has been discussed in detail in (ref. [130]). The question is, does the photon fit into this scheme? Indeed a long time ago, Darwin showed that the massless, force free Dirac theory was formally identical to source free electrodynamics in a vacuum [421]. In the absence of a suitable physical interpretation this mathematical identity has for long been considered to be a mere mathematical coincidence (Cf.ref.[421]). After all, photons are spin one particles, while the Dirac equation represents spin half particles. At the same time, it has also been recognized for a long time– –Einstein and Meyer were one of the first to point this out– –that the spinorial representation of the Lorentz group is more fundamental than the vectorial representation [288]. Indeed an excessive reliance on coincidences is not satisfactory from the point of view of science.

In the light of the above observations we would now like to point out that the above circumstance is not a mere coincidence, but has a definite physical interpretation.

We firstly make some preliminary remarks: Both in electromagnetic theory and in the Dirac theory, the D'Alembertian equation

$$D\psi_\mu = 0 \tag{8.64}$$

where D is the D'Alembertian operator, is satisfied by the respective components. This is merely an expression of Lorentz invariance. At this point the two theories diverge. This is because an equation like (8.64) requires the value of ψ at say $t = 0$ and so also the value of $\frac{\partial\psi}{\partial t}$ for specifying the solution. This does not pose any problem in electromagnetic theory, but is not acceptable in Quantum Theory, because the Quantum Mechanical wave function ψ contains as complete a description of the state as is possible and there is no room for derivatives as initial conditions. This is also the reason why (8.64), or the Quantum Mechanical Klein-Gordan equation gives negative probability densities. So the order of (8.64) needs to be depressed to make it a first order equation, which infact is the starting point of the Dirac theory and leads to the Dirac equation,

$$(\gamma^\mu p_\mu - m)\,\psi = 0 \tag{8.65}$$

It may be mentioned that two component spinors belonging to the representation

$$D^{(\frac{1}{2}0)} \text{or} D^{(0\frac{1}{2})}$$

of the Lorentz group are solutions of the Dirac equation (8.65). But these are no longer invariant under reflections [156]. It is to preserve this invariance that we have to consider the 4×4 representation

$$D^{(\frac{1}{2}0)} \oplus D^{(0\frac{1}{2})}$$

Under reflections, the two spinors transform into each other thus maintaining the overall invariance [399]. We also note that, as is known [7], the Maxwell equations can also be written in the form of neutrino equations. Defining a four vector such that

$$\chi_j = E_j + \imath B_j, \chi_0 = 0 \tag{8.66}$$

we can rewrite the Maxwell equations in the form

$$\beta_\mu \frac{\partial \chi_\nu}{\partial x_\mu} = -\frac{1}{c} j_\nu \tag{8.67}$$

where in a particular representation, for example,

$$\beta_0 = IXI, \quad \beta_1 = -\sigma_3 \otimes \sigma_2,$$

$$\beta_2 = \sigma_2 \otimes I, \quad \beta_3 = \sigma_1 \otimes \sigma_2,$$

the σ's being the Pauli matrices and wherein for our source free vacuum case, the current four vector on the right hand side of equation (8.67) vanishes. It is easy to show that the four component equation (8.67) breaks down into two component neutrino like equations, except that both these equations are coupled owing to the additional condition $\chi_0 = 0$ in (8.66). This has been the problem in identifying (8.67) with the Dirac theory.
In the above context let us now approach the above considerations from the opposite point of view, that of the Dirac equation. It is well known that the four linearly independent four spinor Dirac wave functions are given by [181], apart from multiplicative factors,

$$\begin{bmatrix} 1 \\ 0 \\ \frac{p_z c}{E+mc^2} \\ \frac{p+c}{E+mc^2} \end{bmatrix} \begin{bmatrix} 0 \\ 1 \\ \frac{p-c}{E+mc^2} \\ \frac{-p_z c}{E+mc^2} \end{bmatrix} \begin{bmatrix} \frac{p_z c}{E+mc^2} \\ \frac{p+c}{E+mc^2} \\ 1 \\ 0 \end{bmatrix} \begin{bmatrix} \frac{p+c}{E+mc^2} \\ \frac{-p_z c}{E+mc^2} \\ 0 \\ 1 \end{bmatrix} \tag{8.68}$$

where p_z is the z component of the momentum and

$$p_\pm = p_x \pm \imath p_y,$$

in a representation given by,

$$\gamma_\imath = \gamma_0 \begin{bmatrix} 0 & \sigma_\imath \\ \sigma_\imath & 0 \end{bmatrix}, \gamma_0 = \begin{bmatrix} 1 & 0 \\ 0 & -1 \end{bmatrix}$$

the σ's being the Pauli matrices.
If we consider the z axis to be in the direction of motion, for simplicity and take the limit $m \to 0$, the spinors in (8.68) become,

$$\psi_1 = \begin{bmatrix} 1 \\ 0 \\ 1 \\ 0 \end{bmatrix} \psi_2 = \begin{bmatrix} 0 \\ 1 \\ 0 \\ -1 \end{bmatrix} \psi_3 = \begin{bmatrix} 1 \\ 0 \\ 1 \\ 0 \end{bmatrix} \psi_4 = \begin{bmatrix} 0 \\ -1 \\ 0 \\ 1 \end{bmatrix} \quad (8.69)$$

It should be noticed that in (8.69) $\psi_1 = \psi_3$, and $\psi_2 = \psi_4$ so that effectively, two of the spinors vanishes exactly and we are left with two solutions as in the case of the solutions χ of (8.67).(The mass zero four component Dirac spinor does not represent a neutrino unless an auxiliary condition, which effectively destroys the lower two or upper two components is imposed [399]). It can now be seen from the above considerations that the source free vacuum electromagnetic field can be considered to be a composite of a neutrino and an anti neutrino. We must remember that the equation (8.67) are actually coupled neutrino equations, coupled by the condition in (8.66). It may be mentioned that the possibility of Bosons being bound states of Fermions, rather than being primary has been discussed by the author and other scholars [422, 130].
We now make the following remarks:
1. Already we have referred to Sakharov's formulation of gravitation in terms of the background Zero Point Field (or Quantum vacuum). In this context let us recapitulate the following well known fact encountered in earlier Chapters. Due to the Zero Point oscillators, there is an electromagnetic field density ΔB over an interval L given by

$$(\Delta B)^2 \sim \frac{e^2}{L^4} \quad (8.70)$$

So the energy over an extension $L = l$ is given from (8.70) by $\frac{e^2}{l}$ which is the energy mc^2 of the elementary particle itself,

$$\frac{e^2}{l} = mc^2 \quad (8.71)$$

If on the other hand we replace in (8.71) e^2 by Gm^2, we get, reverting to the length L

$$\frac{Gm^2}{L} \approx mc^2$$

whence

$$L \approx \frac{Gm}{c^2} \quad (8.72)$$

(8.72) shows that as noted earlier, we can similarly obtain from the fluctuating background Zero Point Field a Black Hole, infact a Planck scale Black Hole, it being well known as we saw, that a Planck mass is a Schwarzchild Black Hole at the Planck scale (Cf. also ref.[257]). From this point of view, Planck mass particles (or oscillators) are created from the fluctuation of the Zero Point Field and then lead up to elementary particles as indicated above. In any case, this again brings out the interchangability, $e^2 \to Gm^2$. It is interesting to note that the substitution of $Gm^2 \to e^2 \to g_w^2$ for the neutrino, gives us relations similar to (8.71). That is we get, this time,

$$T = \frac{m\hbar c^3}{8\pi g_w^2} = \frac{m\hbar c^3 \cdot 10^{13}}{8\pi \cdot e^2} = \frac{m_e c^2 \cdot \hbar c \cdot 10^5}{8\pi e^2} \sim 1^\circ,$$

corresponding to the Cosmic Background temperature (of the neutrino background, as well) as we saw, and,

$$l_v = \frac{g_\omega^2}{m_\nu c^2},$$

as already encountered. (Conversely, if we use $T \sim 1^\circ K$, then we recover $g_\omega^2 \sim 10^{-13} e^2$.) The whole point is that as we noted earlier too, there is a complete parallel between the neutrino and an elementary particle which is particularly meaningful in the context of a Planck oscillator underpinning. This can be expressed by,

$$h, m, N, n, T, \bar{N} \to h', m', ', n', 1^\circ K, \bar{N}'$$

where, as we saw in Chapter 6, $h' \sim 10^{-12} h, N'$, the number of neutrinos $\sim 10^{90}, m'$, the neutrino mass is $10^{-10} m, T$ is the Hagedorn temperature and $1^\circ K$ is the corresponding temperature for the neutrino which is the Cosmic Background temperature and which lead to the Fermi temperature considerations earlier, $n' \sim 10^{60}$ is the number of underlying Planck oscillators for a neutrino and $\bar{N}' \sim 10^{125}$ is the number of Planck oscillators providing the underpinning for all the neutrinos. We also note that this cosmic neutrino background was shown to lead to the correct cosmological constant.

Further, the considerations in Section 8.1 showed that Gravitation and Electromagnetism could be thought to be different due to the different "rates" at which these interactions played themselves out. This can be extended to the weak interactions also. The rates are different because of the difference in the number of sub-constituents. All this points to a parallel description of the Universe in terms of the neutrino.

2. We have seen above how from the background Zero Point Field Planck

scale oscillators can "condense". Let us suppose that n such particles are formed. We can then invoke the fact used earlier that [95] for a collection of ultra relativistic particles, in this case the Planck oscillators, the various centres of mass form a two dimensional disk of radius l given by

$$l \approx \frac{\beta}{m_e c} \tag{8.73}$$

where in (8.73) $m_e (\approx m$ in the Large Number sense) is the electron mass and β is the angular momentum of the system. Further l is such that for distances $r < l$, we encounter negative energies (exactly as for the Compton length). It will at once be apparent that for an electron, for which $\beta = \frac{\hbar}{2}$, (8.73) gives the Compton wavelength. We can further characterize (8.73) as follows: By the definition of the angular momentum of the system of Planck particles moving with relativistic speeds, we have

$$\frac{\hbar}{2} = m_P c \int_0^l r^2 dr d\Theta \sim m_P c \sigma l^3 = m_e c l \tag{8.74}$$

In (8.74) we have used the fact that the disk of mass centres is two dimensional, and σ has been inserted to stress the fact that we are dealing with a two dimensional density, so that σ while being unity has the dimension

$$\left[\frac{1}{L^2}\right]$$

The right side of (8.74) gives the angular momentum for the electron. From (8.74) we get

$$\sigma l^2 m_P = m_e \tag{8.75}$$

which ofcourse is correct.
Alternatively from (8.75) we can recover $n \sim 10^{40}$, in the Large Number sense.
In any case, it is amusing to note that, if in (8.75), we replace l by l_ν for the neutrino then we get for the right side, the neutrino mass.

8.9 Quantum Mechanics, General Relativity and The Landscape of Multiply Connected Universes

We start with some comments on Classical Mechanics and Quantum Theory. Though the latter is a radical departure from the former, there is as we saw, a well known formulation that throws up a common denominator, in the Hamilton-Jacobi theory. Moreover General Relativity itself has a

formulation which utilizes the Hamilton-Jacobi theory. We will investigate this commonality first and use the fact that Quantum Theory can be interpreted in a classical context with the additional input of multiply connected spaces, rather than the simply connected space of Classical Theory, as we have already seen [337, 16, 423, 336].
Let us start with the Hamilton-Jacobi theory of Classical Mechanics [209]. The action integral is given by

$$I = \int L\left(\frac{dx}{dt}, x, t\right) dt \tag{8.76}$$

where for the moment, for simplicity we work in one space dimension. This can be easily generalized to three space dimensions. We now extremalize (8.76) [101, 45]. The extremum integral is given by

$$S(x,t) = \bar{I} \tag{8.77}$$

It then follows that [424]

$$p = \frac{\partial S}{\partial x} \tag{8.78}$$

$$E = -\frac{\partial S}{\partial t} = \frac{p^2}{2m} + V(x) \tag{8.79}$$

where p denotes the momentum and E denotes the energy. We can combine these and write,

$$E = H(p,x) \tag{8.80}$$

Using (8.78) and (8.80) in (8.79) we get

$$-\frac{\partial S}{\partial t} = H\left(\frac{\partial S}{\partial x}, x\right) = \frac{1}{2m}\left(\frac{\partial S}{\partial x}\right)^2 + V(x) \tag{8.81}$$

Equation (8.81) is the Hamilton-Jacobi equation in one dimension with a solution given by

$$S = (x,t) = -Et + S\left\{2m\left[E - V\right]\right\}^{1/2} dx + S \tag{8.82}$$

To facilitate later work, we re-introduce a wave mechanical description

$$\psi = Re^{(\imath/\hbar)S(x,t)} \tag{8.83}$$

where R is slowly varying and S is given by (8.77). Equations (8.81) and (8.82) can now be given a more general character in terms of S, which is the so called dynamical phase. Let us now impose the requirement

$$\frac{\partial S}{\partial E} = 0 \tag{8.84}$$

which is now interpreted as a constructive interference in the phase of systems described by wave functions (8.83). Then as is well known (Cf.[101]) we are lead back to a description from a wave to a wave packet or particle description as in (8.79) or (8.81).

The whole point of this discussion is to show the possible commonality with Quantum Theory on the one hand and Classical Mechanics on the other. We will now proceed further in this direction.

Before proceeding to the nuances of Quantum Theory vis-a-vis Classical Mechanics, we would like to point out that a very similar line of reasoning can also be applied to General Relativity, as shown by Wheeler and others a long time ago (Cf. for example [101, 45]), except that the arena is now superspace, which is a manifold in which a single point represents the whole geometry of three dimensional space.

In this development we denote a three dimensional geometry by $G^{(3)}$. This plays the role of spacetime coordinates in Classical Mechanics. Treating this as a point we introduce the wave function like description (8.83), except that we are in superspace which is essentially a four dimensional manifold. The four Geometry $G^{(4)}$ is now the analogue of the history or trajectory of a particle. Interestingly it has been shown by Stern that as long as we are dealing with Euclidean type three dimensional spaces with positive definite metric, superspace constitutes a manifold in the sense that each point therein has a neighborhood homeomorphic to an open set in a Banach space and two distinct points are separated by disjoint neighborhoods. This enables us to carry out the usual operations in superspace and we are led back using the principle of constructive interference (8.84) to this time the ten field equations of Einstein (Cf.ref.[101] for details).

Thus the Hamilton-Jacobi theory leads both to Quantum Mechanics and to General Relativity in terms of the wave function description (8.83) together with constructive interference. However it must be borne in mind that in Quantum Mechanics we are dealing with the usual three dimensional space whereas for obtaining Einstein's equations of General Relativity, we are using the four dimensional superspace.

Let us now return to Quantum Theory. To re-emphasize we start by reviewing Dirac's original derivation of the monopole though this time our motivation is different. He started with the wave function similar to (8.83):

$$\psi = Ae^{\imath\gamma}, \tag{8.85}$$

It will be recalled that he then considered the case where the phase γ in (8.85) is non integrable. In this case (8.85) can be rewritten as

$$\psi = \psi_1 e^{\imath S}, \qquad (8.86)$$

where ψ_1 is an ordinary wave function with integrable phase, and further, as we saw, while the phase S does not have a definite value at each point, its four gradient viz.,

$$K^\mu = \partial^\mu S \qquad (8.87)$$

is well defined. We use temporarily natural units, $\hbar = c = 1$. Dirac then went on to identify K in (8.87) (except for the numerical factor hc/e) with the electromagnetic field potential, as in the Weyl gauge invariant theory. Next we considered the case of a nodal singularity, which as we saw is closely related to what was later called a quantized vortex (Cf. for example ref.[297]). In this case a circuit integral of a vector as in (8.87) gives, in addition to the electromagnetic term, a term like $2\pi n$, so that we have for a change in phase for a small closed curve around this nodal singularity,

$$2\pi n + e \int \vec{B} \cdot d\vec{S} \qquad (8.88)$$

In (8.88) $\vec{B}$ is the magnetic flux across a surface element $d\vec{S}$ and n is the number of nodes within the circuit. The expression (8.88) directly lead to the Monopole in Dirac's formulation.

With the above background we now consider the universe at large. As we saw the Hamilton-Jacobi theory leads both to Quantum Mechanics and to General Relativity in terms of the wave function description (8.83) together with constructive interference.

To push these considerations even further let us re-consider the general relativistic formulation in terms of linearized theory. In fact this can be applied to the case of the electron, and it was shown [130] that the spin was given by,

$$S_K = \int \epsilon_{klm} x^l T^{m0} d^3x = \frac{h}{2} \qquad (8.89)$$

where the domain of integration was a sphere of radius given by the Compton wavelength. If this is carried over to the case of the Universe, it has been shown that we get from (8.89)

$$S_U = N^{3/2} h \approx h_1 \qquad (8.90)$$

where h_1 which is given by (8.1) and S_U denotes the counterpart of electron spin and $N \sim 10^{80}$ is the number of elementary particles in the Universe.

h_1 in (8.90) turns out to be the spin of the universe itself as noted earlier in broad agreement with Godel's spin value for Einstein's equations [425, 408]. Incidentally this is also in agreement with the Kerr limit of the spin of the rotating Black Hole. Further as pointed out by Kogut and others, the angular momentum of the universe given in (8.90) is compatible with a rotation from the cosmic background radiation anisotropy [408]. Finally it is also close to the observed rotation as deduced from anisotropy of cosmic electromagnetic radiation as reported by Nodland and Ralston and others [426, 427].

As we saw $h_1 \sim 10^{93}$ is the large scale analogue of the Planck constant and we immediately got, in support of this analogy, equation (8.10).

Thus, using the formulation of linearized General Relativity as in (8.89) and (8.90) we can get a unified description of Quantum Mechanical spin, that is equation (8.89), and the Universe itself in a similar description as in (8.90) and (8.1). What we are saying, as we will stress again later, is that the Universe is rather like a blown up version of an electron.

All this becomes relevant in the light of recent developments. It is now generally believed that our observable universe is but one amongst a very large number of what may be roughly called parallel universes, or what in the context of Super String theory are called multiverses. There are different routes to this conclusion. If one goes by String theory, then there are 10^{500} different solutions, each corresponding to a different universe. In any case all this is in the spirit of "Scaled" Quantum Mechanics--we need not stop at the Universe as we know it--this may be just another scale or step. We will come back to this theme in a moment.

Another line of reasoning is that inflation has created any number of bubble type universes, our own universe being one such bubble. From yet another point of view, many astronomers are now coming round to accepting the many worlds interpretation of Hugh Everett III--each of the Quantum possibilities defines a different universe, unlike the Copenhagen interpretation in which there is an acausal collapse of the wave function into a definite state. Yet another route to the many universe model is through the singularity thrown up by Einstein's theory of General Relativity. The singularity itself defines any number of different universes with different laws of physics. We have already noted that the observable universe is a giant black hole, a blown up version of an elementary particle. There can be any number of such universes [428, 424]. Indeed we know that

$$R \approx \frac{2GM}{c^2},$$

that is the Universe mimics a Schwarzchild Black Hole. (If the latest WMAP and other results are confirmed, in addition to the usual mass M, we have to consider the dark matter mass as well, which may be some six times greater.) In any case an n(space) dimensional black hole is embedded in at least $(n+1)$ space dimensions.

Whatever be the route to the conclusion, we can now take the multiply connected space considerations of micro physics which we saw above to the case of the macro cosmos. In this case super space would be multiply connected consisting of three (space) dimensional universes rather than points $G^{(3)} \sim s$ in the Wheeler formulation, which mimic particles with spin. In other words, the different "parallel" universes would be multiply connected and they would be embedded in a higher dimensional superspace. A rough rationale for the higher dimensionality (a la Wheeler's 4-space geometry) is the following: If we imagine our universe of radius R to be a point, then a sequence of points of zero dimension each, form a line of one dimension, a sequence of lines would form a two dimensional surface and so on.

This process could go on, as in the multi verse formulation.

To proceed further, we first note that in Section 8.1, we have already characterized the Universe as a Gaussian wave packet. We also saw in Section 8.4, that different mass concentrations like clusters of galaxies in the Universe can be characterized like "energy levels" in (8.45) extending this to the Universe itself with h_1 given by (8.1), we get

$$\frac{GM^3R}{h_1^2} \sim 10,$$

which consistently gives for R, the radius of the Universe, if M is its mass. In other words, the Universe itself is an "energy level".

To press these considerations further we note that we can use the Beckenstein temperature formula for a black hole

$$T = \frac{hc^3}{8\pi Gkm}$$

replacing h by h_1 and the mass m by M, the mass of the universe to get

$$kT = Mc^2$$

This is consistent, as the right side is the energy content of the universe. Further the decay time in this black hole is given by, as we saw

$$t = \frac{1}{3\beta}m^3, \; \beta = \frac{hc^4}{(30.8)^3\pi G^2}$$

With the scaled up Planck constant and the mass of the Universe we get, the age to be

$$t \sim 10^{21} s$$

This shows that the universe would decay in a time span that is $\sim 10^4$ times its present age. Let us also generalize our formula for the gravitational constant,

$$G = \frac{lc^2}{\sqrt{N}m} = \frac{l^2 c}{mt}$$

We can then speak of the force between the different universes within the super universe. If we take the number of universes in the super universe to be $\sim 10^{100}$, say, then the replacement in the above with the cosmic parameters gives the analogue of the gravitational constant as

$$G' \sim 10^{-55}$$

This constant is some 10^{-47} times as weak as the gravitational constant itself.

8.10 The Monopole

Joshi and Ignatieu had shown that a non zero photon mass implies the non existence of the monopole [295]. We on the other hand have argued that the photon indeed has a mass. So, there should not be any monopoles and indeed they have not been found, as noted. This is what we hade shown some years ago [297] and will argue now.
Ever since Dirac deduced theoretically the existence of the monopole in 1931, it has eluded physicists [51]. At the same time the possibility of realising huge amounts of energy using monopoles has been an exciting prospect. In 1980 when the fiftieth Anniversary of the monopole was being commemorated, Dirac himself expressed his belief that the monopole did not exist [298]. Some scholars have indeed dismissed the monopole [429, 430], while in a model based on quantized vortices in the hydrodynamical formulation, the monopole field can be mathematically identified with the momentum vector [129]. Monopoles had also been identified with solitons [431].
In any case, it has been noted that the existence of free monopoles would lead to an unacceptably high density of the universe [43], which in the light of latest observations of an ever expanding universe [177, 432] would be

difficult to reconcile.

We will now show that monopoles arise due to the non commutative structure of space time being ignored, and this would also provide an explanation for their being undetected.

Let us start by reviewing Dirac's original derivation of the Monopole (Cf.ref.[51]). He started with the wave function in equation (8.85). We can then go on as in the previous section till equation (8.88).

In (8.88) $\vec{B}$ is the magnetic flux across a surface element $d\vec{S}$ and n is the number of nodes within the circuit. The expression (8.88) directly leads to the Monopole.

Let us now reconsider the above arguments in terms of our non commutative geometry [297].

As noted above, the non integrability of the phase S in (8.86) gives rise to the electromagnetic field, while the nodal singularity gives rise to a term which is an integral multiple of 2π. As is well known [127] we have

$$\vec{\nabla} S = \vec{p} \tag{8.91}$$

where $\vec{p}$ is the momentum vector. When there is a nodal singularity, as noted above the integral over a closed circuit of $\vec{p}$ does not vanish. In fact in this case we have a circulation given by

$$\Gamma = \oint \vec{\nabla} S \cdot d\vec{r} = \hbar \oint dS = 2\pi n \tag{8.92}$$

It is because of the nodal singularity that though the $\vec{p}$ field is irrotational, there is a vortex−−the singularity at the central point associated with the vortex makes the region multiply connected, or alternatively, in this region we cannot shrink a closed smooth curve about the point to that point. Infact if we use the fact as seen above that the Compton wavelength is a minimum cut off, then we get from (8.92) using (8.91), and on taking $n = 1$,

$$\oint \vec{\nabla} S \cdot d\vec{r} = \int \vec{p} \cdot d\vec{r} = 2\pi mc \frac{l}{2mc} = \frac{h}{2} \tag{8.93}$$

$l = \frac{\hbar}{2mc}$ is the radius of the circuit and $h = 2\pi$ in the above natural units. In other words the nodal singularity or quantized vortex gives us the mysterious Quantum Mechanical spin half (and other higher spins for other values of n). In the case of the Quantum Mechanical spin, there are $2 \times n/2 + 1 = n + 1$ multiply connected regions, exactly as in the case of nodal singularities. Indeed in the case of the Dirac wave function, which is a bi-spinor $\begin{pmatrix} \Theta \\ \phi \end{pmatrix}$, we have noted that the double connectivity was shown to lead immediately to the same electromagnetic potential we had obtained from the nonintegrability of the phase above, which again was identical to that from Weyl's guage invariant theory.

Let us revisit all this in a little greater detail. We start with a non integrable infinitessimal parallel displacement of a four vector [39],

$$\delta a^\sigma = -\Gamma^\sigma_{\mu\nu} a^\mu dx^\nu \tag{8.94}$$

The Γ's are the Christoffel symbols. This represents the extra effect in displacements, due to curvature. In a flat space, all the Γ's on the right side would vanish. Considering partial derivatives with respect to the μ-th coordinate, this would mean that, due to (8.94), as we saw,

$$\frac{\partial a^\sigma}{\partial x^\mu} \to \frac{\partial a^\sigma}{\partial x^\mu} - \Gamma^\sigma_{\mu\nu} a^\nu, \tag{8.95}$$

The second term on the right side of (8.95) was shown to lead to

$$\frac{\partial}{\partial x^\mu} \to \frac{\partial}{\partial x^\mu} - \Gamma^\nu_{\mu\nu} \tag{8.96}$$

We identified

$$A_\mu = \Gamma^\nu_{\mu\nu} \tag{8.97}$$

from the above using minimum electromagnetic coupling exactly as in Dirac's monopole theory.

If we use (8.96), we will get the commutator relation,

$$\frac{\partial}{\partial x^\lambda}\frac{\partial}{\partial x^\mu} - \frac{\partial}{\partial x^\mu}\frac{\partial}{\partial x^\lambda} \to \frac{\partial}{\partial x^\lambda}\Gamma^\nu_{\mu\nu} - \frac{\partial}{\partial x^\mu}\Gamma^\nu_{\lambda\nu} \tag{8.98}$$

Let us now use (8.97) in (8.98): The right side does not vanish due to the electromagnetic field (8.97) and we have a non-commutativity of the momentum components of quantum theory. Indeed the left side of (8.98) can be written as

$$[p_\lambda, p_\mu] \approx \frac{0(1)}{l^2}, \tag{8.99}$$

l being the Compton wavelength. In (8.99) we have utilized the fact that at the extreme scale of the Compton wavelength, the Planck scale being a special case, the momentum is mc.

From (8.97), (8.98) and (8.99), we have,

$$Bl^2 \sim \frac{1}{e} = \left(\frac{\hbar c}{e}\right), \tag{8.100}$$

where B is the magnetic field.

Equation (8.100) is the well-known equation for the magnetic monopole.

Indeed it has been shown by Saito and the author [371, 372] that a non commutative spacetime at the extreme scale shows up as a powerful magnetic field.
To recapitulate, the Monopole was shown by Dirac to arise because of two separate issues. The first was the non integrability of the phase S given in (8.86), which gave rise to the electromagnetic potential (8.87) which was equivalent to the Weyl potential (8.97) (which latter was originally dismissed because it was adhoc). The other issue was that of nodal singularities or alternatively the multiply connected nature of space which gave rise to a term like $2\pi n$ as in (8.88). In effect there would be free monopoles. However all this was considered in the context of the usual commutative Minkowski spacetime. Effectively this means that terms $\sim 0(l^2)$ are neglected.
However once such terms are included, in other words once the non commutative (and multiply connected) structure of spacetime to this order is recognized, firstly the previously supposedly adhoc Weyl electromagnetic formulation automatically follows and furthermore the first term in the monopole expression (8.88) immediately gives the Quantum Mechanical spin, and the elusive monopole appears as the magnetic effect at the Compton (or Planck scale).

8.11 Fermions and Bosons

We first observe that a hydrodynamical vortex and streamline description provides an explanation for the Fermionic and Bosonic statistics [130]. Indeed, let n_K be the occupation number for the energy or momentum state defined by K. For Fermions $n_K = 0$ or 1, whereas n_K can be arbitrary for Bosons. We can give a rationale for this difference between Bose-Einstein and Fermi-Dirac statistics. Fermions are bounded by the Compton wavelength. That is, they are localized, a description which, as we have seen, requires both negative and positive energy solutions, which in fact is expressed by zitterbewegung effects. The localization in space automatically implies an indeterminacy of energy or momentum K. Thus, even though an energy or momentum state, in practical terms implies a small spread ΔK, it is not possible to cram Fermions which also have a momentum energy indeterminacy spread, arbitrarily into this state.
On the other hand Bosons are not bound by the Compton wavelength vortices, and so have sharper momentum states, so that any number of them

can be crammed into the state K (which really is blurred by the Uncertainty Principle indeterminacy of ΔK). This point will be seen clearly in the sequel. Another well known way of expressing these facts is by saying that the fermionic wave function in space is weak, but not the bosonic wave function, the latter fact being symptomatic of a field or an interaction[319]. The above considerations immediately follow from our description in terms of fuzzy spacetime. In this case we have the non commutative geometry given earlier

$$[x, y] = 0(l^2), [x, p_x] = \imath\hbar[1 + l^2] \tag{8.101}$$

where l is the Compton wavelength. As we saw in Chapter 6, it is precisely this space quantization at the Compton scale that leads to the Dirac matrices and their anti-commutation relations.

Bosons on the other hand would be described as will be seen below, and as mentioned earlier as bound states of the Fermions (Cf.[422] and other references), or alternatively they would be a super position of vortices leading to a streamline like description [433] rather than be tight quantized vortices. Indeed it has been noted that the spinorial representation of the Lorentz group is more fundamental than the vectorial representation[143].

If indeed the Fermion-Boson divide is not so rigid then, within the above description there could be certain special situations in which the above space localized and momentum space localized description of Fermions and Bosons gets blurred, in which case anomalous or fractal statistics would come into play. This could happen, for example when the Compton wavelength l of the Fermion becomes very small, that is the particle is very massive. In this case the non commutativity of the geometry referred to above in (8.101) disappears and we return to the usual commutation relations of non relativistic Quantum Mechanics, that is a description in terms of the spinless Schrodinger equation. In this case, v being the velocity of the particle v/c would be small and the Dirac equation as is known tends to the Schrodinger equation[181].

There would thus be a Bosonisation effect. This would also be expected at very low temperatures, for example below the Fermi temperature, when the energy spread of the Fermions would itself be small (or the Compton scale would be small in comparison to scales of interest in the problem), and anomalous behaviour, for example on the lines of the superfluidity of He^3[248, 434] can be expected.

This Bosonization effect is also suggested by the following argument which we saw earlier in a slightly different context:

For a collection of Fermions, we know that the Fermi energy is given by [125],

$$\epsilon_F = p_F^2/2m = (\frac{\hbar^2}{2m})(\frac{6\pi^2}{v})^{2/3} \tag{8.102}$$

where $v^{1/3}$ is the interparticle distance. On the other hand, in a different context, for phonons, as we saw the maximum frequency is given by, (cf.ref.[125]),

$$\omega_m = c(\frac{6\pi^2}{v})^{1/3} \tag{8.103}$$

This occurs for the phononic wavelength $\lambda_m \approx$ inter-atomic distance between the atoms, $v^{1/3}$ being, again, the mean distance between the phonons. "c" in (8.103) is as we saw the velocity of the wave, the velocity of sound in this case. The wavelength λ_m is given by,

$$\lambda_m = \frac{2\pi c}{\omega_m}$$

We can now define the momentum p_m via the de Broglie relation,

$$\lambda_m = \frac{h}{p_m},$$

which gives,

$$p_m = \frac{\hbar}{c}\omega_m, \hbar \equiv \frac{h}{2\pi} \tag{8.104}$$

We can next get the maximum energy corresponding to the maximum frequency ω_m given by (8.103),

$$\epsilon_m = \frac{p_m^2}{2m} = \frac{\hbar^2}{2m}(\frac{6\pi^2}{v})^{2/3} \tag{8.105}$$

Comparing (8.102) and (8.105), we can see that ϵ_m and p_m exactly correspond to ϵ_F and p_F.

The Fermi energy in (8.102) is obtained as is known by counting all energy levels below the Fermi energy ϵ_F using Fermi-Dirac statistics, while the maximum energy in (8.105) is obtained by counting all energy levels below the maximum value, but by using Bose-Einstein statistics.

We can see why in spite of this, the same result is obtained in both cases. In the case of the Fermi energy, all the lowest energy levels below ϵ_F are occupied with the fermionic occupation number $\langle n_p \rangle = 1, p < p_F$. Then, the number of levels in a small volume about p is d^3p. This is exactly so for the bosonic levels also. With the correspondence given in (8.104), the number of states in both cases coincide and it is not surprising that (8.102)

and (8.105) are the same.
In effect, Fermions below the Fermi energy should have a strong resemblance to phonons, reminiscent of semions which behave like particles with statistics in between the Fermi-Dirac and Bose-Einstein statistics [435].
A rationale for the above is the fact that, for $\frac{p^2}{2m} < \epsilon_F$, as $\Delta p \approx 0$, the levels are very closely spaced and the density of levels in p-space is d^3p as in the case of phonons which are Bosons. This illustrates comments made earlier about momentum states and occupation members. The point is that Fermions are bounded by the Compton wavelength, as we have stressed repeatedly. So they do not have sharp momentum states and the occupation number n_k (in a small momentum interval $(k - \Delta k/2, k + \Delta k/2)$) cannot be arbitrarily large, but rather $n_k = 1 \,\text{or}\, 0$. Bosons on the other hand have sharp momentum states and n_k the occupation number in the small momentum interval can be arbitrary. Bloch's original analysis (cf.ref.[436]) corroborates the above considerations.
In any case, for example the conduction electrons in metals can, in the Sommerfeld model be considered to be non interacting Fermions in a box, as also in the Landau theory of Fermi liquids [437]. Moreover the original Tomonaga theory in one dimension which considers the fermionic ensemble as an ensemble of Bosons and weakly or non interacting Fermions, as in the independent particle model has been found to be true in three dimensions also [436]. It is in this context that we can use the above result that the Dirac equation goes over to the Schrodinger equation at low velocities, to argue that there would be Bosonization effects. So at sufficiently low velocities, that is temperatures, we can expect that these Fermions would exhibit a bosonic character.
Let us analyse this circumstance further to show that the above conclusions indeed follow from Quantum Field Theory, without any contradiction to the Spin Statistics Theory.
We first show specifically that the Fermi energy corresponds to scales much larger than the Compton wavelength[248]. This follows quite easily. If v is the average volume per particle, then scales much larger than the Compton wavelength imply,

$$v >> (\frac{\hbar}{mc})^3,$$

whence, in terms of the Fermi energy, we have (cf.ref.[125])

$$(\frac{3\sqrt{\pi}}{4})^{2/3} . \frac{1}{kT\epsilon_F}(\frac{2\pi\hbar^2}{mkT}) >> (\frac{\hbar}{mc})^2$$

So,

$$(7.6)mc^2 >> \epsilon_F$$

Alternatively,

$$\epsilon_F \sim \frac{1}{N} \sum \frac{p^2}{2m} << mc^2,$$

$$\text{as} \quad (\frac{p}{mc}) << 1.$$

Either way, the length scales associated with the Fermi energy are much greater than the Compton wavelength.
We next consider these results in the context of the Spin Statistics Theory. We observe that the essence of Spin Statistics Theory is that commutators (corresponding to symmetric wave functions) cannot be used with Fermionic fields while anti commutators (corresponding to anti symmetric wave functions), cannot be used with Bosonic fields. However, as is known, this is strictly true at scales not much greater than the Compton wavelength [438].
Indeed, for a Klein-Gordon field while the vaccum expectation value of the commutator,

$$< 0|[\phi_r(x), \phi_s(y)]|0 >\equiv \Delta(x-y)$$

vanishes for space like intervals, the same value for the anti commutator for large spatial distances is given by,

$$\Delta_1'(x,0) \sim \frac{Ze^{-m|x|}}{|x|^2} + \int_{m_1^2}^{\infty} d\sigma^2 \rho(\sigma^2) \frac{e^{-\sigma|x|}}{|x|^2}$$

So this anti commutator is nearly zero for large space like distances, that is the violation of microscopic causality and therefore the Spin Statistics Theory is negligible (cf.ref.[438]).
Similarly for Fermionic fields the contradiction arises because, this time the symmetric propogator, the Lorentz Invariant function

$$\Delta(x-x') \equiv \int \frac{d^3k}{(2\pi)^3 3\omega_k} [e^{-\imath k.(x-x')} + e^{\imath k.(x-x')}]$$

does not vanish for space like intervals $(x-x')^2 < 0$, where the vacuum expectation value of the commutator is given by the spectral representation,

$$S(x-x') \equiv \imath < 0|[\psi_\alpha(x), \psi_\beta(x')]|0 >$$

$$= - \int dM^2 [\imath \rho_1(M^2) \Delta_x + \rho_2(M^2)]_{\alpha\beta} \Delta(x-x')$$

Outside the light cone, $r > |t|$, where $r \equiv |\vec{x} - \vec{x}'|$ and $t \equiv |x_o - x'_o|, \Delta$ is given by,

$$\Delta(x' - x) = -\frac{1}{2\pi^2 r}\frac{\partial}{\partial r}K_o(m\sqrt{r^2 - t^2}),$$

where the modified Bessel function of the second kind, K_o is given by,

$$K_o(mx) = \int_o^\infty \frac{\cos(xy)}{\sqrt{m^2 + y^2}}dy = \frac{1}{2}\int_{-\infty}^\infty \frac{\cos(xy)}{\sqrt{m^2 + y^2}}dy$$

(cf.[439, 440]). In our case, $x \equiv \sqrt{r^2 - t^2}$, and we have,

$$\Delta(x - x') = const\frac{1}{x}\int_{-\infty}^\infty \frac{y \sin xy}{\sqrt{m^2 + y^2}}dy \sim 0(\frac{l}{x}),$$

where l is the Compton wavelength ($\hbar = c = 1$).
Once again we can see that the violation of the Spin Statistics Theory is negligible for distances large compared to the Compton wavelength. This confirms the Bosonization effects for low temperature Fermions.
To get yet another alternative justification, we further observe that in the Quantum Field Theory of Fermions, as is well known (cf.ref.[438]), the wave function expansion of the Fermion should include solutions of both signs of energy:

$$\psi(\vec{x}, t) = N \int d^3p \sum_{\pm s} [b(p, s)u(p, s) \exp(-\imath p^\mu x_\mu/\hbar) + d^*(p, s)v(p, s) \exp(+\imath p^\mu x_\mu/\hbar)] \quad (8.106)$$

where N is a normalization constant for ensuring unit probability.
In Quantum Field Theory, the coefficients in (8.106) become creation and annihilation operators while bb^+ and d^+d become the particle number operators with eigen values 1 or 0 only. The Hamiltonian is now given by:

$$H = \sum_{\pm s} \int d^3pE_p[b^+(p, s)b(p, s) - d(p, s)d^+(p, s)] \quad (8.107)$$

As can be seen from (8.107), the Hamiltonian is not positive definite and it is this circumstance which necessitates the Fermi-Dirac statistics.
Now in our fuzzy spacetime model all the negative energies are pinched off inside the Compton wavelength. So, at the scales under consideration, these are inaccessible, so that there is no question of transition to a negative energy level. That is, we do not require Fermi-Dirac statistics, which was invoked only to forbid such a transition. In effect we could work with commutators. This is reminiscent of the Bosonization of Fermions encountered

in the one dimensional case (cf.ref.[436]).
However it must be observed that the above anomalous behaviour does not mean that the gas behaves classically or that the Pauli Exclusive Principle is inoperative. If that were the case the internal energy at the temperature $T \approx 0°K$ would have been zero, but owing to the existence of the Fermi energy ϵ_F, this internal energy density which is proportional to ϵ_F, is non zero [441].
For very light Fermions, for example neutrinos, the Compton wavelength becomes very large, and in this case the double connectivity of spacetime all but disappears and the observed anomalous features of the neutrino like handedness show up, as discussed elsewhere in detail [218, 442].
Moreover with recent developments in nano technology and thin films, we are able to consider one dimensional and two dimensional Fermions, in which case, Bosonisation effects show up as discussed below. In any case in the one dimensional and two dimensional cases, the Dirac equation becomes a two component equation, without an invariant mass[46], while at the same time, handedness shows up [156]. It is worth mentioning here that the Dirac matrices can have only even dimensionality, corresponding to the above anomalous two component Dirac spinors, and the usual Dirac bispinors of the three dimensional theory.
We very briefly comment on what happens in the two and one dimensional cases in the context of the considerations seen above. These are two extreme idealizations because as we have noted earlier it is spin half that leads to and is responsible for three dimensions. Side stepping this issue for the moment and also the fact that this corresponds to constrained Quantum systems, we observe that this is opposite to the previous situation. We are in the high energy relativistic domain in the sense that the shrinkage of even a single dimension implies that we are already at the Compton wavelength, and the concept of the particle inertial mass and other properties become questionable.
In this case, we encounter mostly the negative energy components which as we have seen exhibit the lefthanded behaviour as in the case of the neutrino too. We can now argue, exactly as we did for the QFT Hamiltonian (8.107), that the question of transition to empty states of the Dirac sea of opposite sign of energy does not arise as these states are unavailable. Whence we can use commutators instead of anti commutators. These conclusions can be easily verified.
To further clarify this situation and demonstrate self consistency within our model let us take the Lorentz covariant equation in one (spatial) dimension,

in a well known and obvious notation:

$$(\imath^\mu \gamma^\mu \partial_\mu - \frac{mc}{\hbar})\psi = 0$$

Apart from the fact that we get the left handed solution, it must be noticed that the mass term (or the energy operator term) is not accompanied by the usual factor,

$$\begin{pmatrix} 1 & 0 \\ 0 & -1 \end{pmatrix}$$

which in fact gives the positive and negative energy solutions and the zitterbewegung, and which leads to equations like (8.106) and (8.107), or to the Fermion bounded by the Compton wavelength, and inertial mass itself as seen in Chapter 2.

Another way of looking at this is that if we work only with solutions of one sign, the current, or equivalently, the expectation value of the velocity operator $c\vec{\alpha}$, is given by (cf.ref.[181]),

$$J^+ = \langle c\vec{\alpha}\rangle_+ = \langle c^2 p/E\rangle_+ = \langle v_{gp}\rangle_+,$$

which is a contradiction, because, $c\vec{\alpha}$ has eigen values $\pm c$, whereas we require $\langle v_{gp}\rangle < c$, if the particle has mass. So, either the particle has no mass, as we saw for the neutrino, or both positive and negative energy solutions have to be included.

In our case, we have neutrino like particles.

Indeed in low dimensions we have Fermion-Boson Transmutation and other statistics like anyonic statistics[435, 436]. We can in fact show that the assembly behaves as if it is at a temperature below the Fermi Temperature: The average energy per unit length in one dimension is given by

$$e = \frac{\pi(kT)^2}{6\hbar\nu_F} \tag{8.108}$$

where $\nu_F \equiv \hbar\pi(N/L)/m, L$ being the length of the one dimensional wire and N the number of Fermions therein. This is the one dimensional version of the Stephan Boltzmann law for radiation[108]. Denoting the average interparticle distance,

$$\frac{L}{N} \equiv (\nu)^{1/3},$$

and using the fact that [125]

$$kT_F = (\frac{\hbar^2}{2m})(\frac{6\pi^2}{\nu})^{2/3},$$

and remembering that,

$$kT = e\nu^{1/3},$$

we can easily deduce from (8.108) that,

$$T = \frac{3}{5} T_F$$

Interestingly this not only shows that the temperature is below the Fermi temperature, but also that the gas is in the ground state [125], whatever be the temperature.

We now consider in greater detail two illustrative situations where anomalous behaviour shows up[443–445].

A. Nearly Mono Energetic Fermions

In this case, the earlier comments lead us to expect Bosonic behaviour because there is hardly any energy spread. Our starting point is the well known formula for the occupation number of a Fermion gas[125]

$$\bar{n}_p = \frac{1}{z^{-1}e^{bE_p} + 1} \tag{8.109}$$

where, $z' \equiv \frac{\lambda^3}{v} \equiv \mu z \approx z$ because, here, as can be easily shown $\mu \approx 1$,

$$v = \frac{V}{N}, \lambda = \sqrt{\frac{2\pi\hbar^2}{m/b}}$$

$$b \equiv \left(\frac{1}{KT}\right), \quad \text{and} \quad \sum \bar{n}_p = N \tag{8.110}$$

Let us consider in particular a collection of Fermions which is somehow made nearly mono-energetic, that is, given by the distribution,

$$n'_p = \delta(p - p_0)\bar{n}_p \tag{8.111}$$

where $\bar{n}_p$ is given by (8.109).

This is not possible in general——here we consider a special situation of a collection of mono-energetic particles in equilibrium which is the idealization of a contrived experimental set up.

By the usual formulation we have,

$$N = \frac{V}{\hbar^3}\int d\vec{p}n'_p = \frac{V}{\hbar^3}\int \delta(p - p_0)4\pi p^2 \bar{n}_p dp = \frac{4\pi V}{\hbar^3}p_0^2 \frac{1}{z^{-1}e^{\theta} + 1} \tag{8.112}$$

where $\theta \equiv bE_{p_0}$.

It must be noted that in (8.112) there is a loss of dimension in momentum space, due to the δ function in (8.111)−−in fact such a fractal two dimensional situation would in the relativistic case lead us back to the anomalous behaviour already alluded to [446]. This again is symptomatic of distances in space (and momentum space) being, more a measure of dispersion, rather than rigid distances, as discussed in Chapter 2. In the non relativistic case two dimensions would imply that the coordinate ψ of the spherical polar coordinates (r, ψ, ϕ) would become constant, $\pi/2$ in fact. In this case the usual Quantum numbers l and m of the spherical harmonics [144] no longer play a role in the usual radial wave equation

$$\frac{d^2u}{dr^2} + \left\{ \frac{2m}{\hbar^2}[E - V(r)] - \frac{l(l+1)}{r^2} \right\} u = 0, \tag{8.113}$$

The coefficient of the centrifugal term $l(l+1)$ in (8.113) is replaced by m^2 as in Classical Theory[209].
To proceed, in this case, $KT = < E_p > \approx E_p$ so that, $\theta \approx 1$. But we can continue without giving θ any specific value.
Using the expressions for v and z given in (8.110) in (8.111), we get

$$(z^{-1}e^\theta + 1) = (4\pi)^{5/2}\frac{z'^{-1}}{p_0}; \text{whence}$$

$$z'^{-1}A \equiv z'^{-1}\left(\frac{(4\pi)^{5/2}}{p_0} - e^\theta\right) = 1, \tag{8.114}$$

where we use the fact that in (8.110), $\mu \approx 1$ as can be easily deduced.
A number of conclusions can be drawn from (8.114). For example, if,

$$A \approx 1, i.e.,$$

$$p_0 \approx \frac{(4\pi)^{5/2}}{1+e} \tag{8.115}$$

where A is given in (8.114), then $z' \approx 1$. Remembering that in (8.110), λ is of the order of the de Broglie wave length and v is the average volume occupied per particle, this means that the gas gets very densely packed for momenta given by (8.115). Infact for a Bose gas, as is well known, this is the condition for Bose-Einstein condensation at the level $p = 0$ (cf.ref.[125]).
On the other hand, if,

$$A \approx 0(\text{that} \quad \text{is} \quad \frac{(4\pi)^{5/2}}{e} \approx p_0)$$

then $z' \approx 0$. That is, the gas becomes dilute, or V increases.
More generally, equation (8.114) also puts a restriction on the energy (or momentum), because $z' > 0$, viz.,

$$A > 0 (i.e. p_0 < \frac{(4\pi)^{5/2}}{e})$$

$$\text{But} \quad \text{if} A < 0, (i.e. p_0 > \frac{(4\pi)^{5/2}}{e})$$

then there is an apparent contradiction.
The contradiction disappears if we realize that $A \approx 0$, or

$$p_0 = \frac{(4\pi)^{5/2}}{e} \tag{8.116}$$

(corresponding to a temperature given by $KT = \frac{p_0^2}{2m}$) is a threshold momentum (phase transition). For momenta greater than the threshold given by (8.116), the collection of Fermions behaves like Bosons. In this case, the occupation number is given by

$$\bar{n}_p = \frac{1}{z^{-1}e^{bE_p} - 1},$$

instead of (8.109), and the right side equation of (8.114) would be given by $' - 1'$ instead of $+1$, so that there would be no contradiction. Thus in this case there is an anomalous behaviour of the Fermions.

B. Degenerate Bosons

We could consider a similar situation for Bosons also where an equation like (8.111) holds. In this case we have equations like (8.115) and (8.116):

$$p_0 \approx \frac{(4\pi)^{5/2}}{1.4e - 1} \tag{8.117}$$

$$p_0 \approx \frac{(4\pi)^{5/2}}{e} \tag{8.118}$$

(8.118) is the same as (8.116), quite expectedly. It gives the divide between the Fermionic and Bosonic behaviour in the spirit of the earlier remarks. At the momentum given by (8.117) we have a densely packed Boson gas rather as in the case of Bose Einstein condensation. On the other hand at the momentum given by (8.118) we have infinite dilution, while at lower momenta than in (8.118) there is an anomalous Fermionisation.
Finally it may be pointed out that at very high temperatures, once again the

energy−−momentum spread of a Bosonic gas becomes large, and Fermionisation can be expected, as indeed has been shown elsewhere [447]. In any case at these very high temperatures, we approach the Classical Maxwell Boltzmann situation.

To sum up Fermions and Bosons are divided into two different compartments, obeying Fermi-Dirac and Bose-Einstein statistics respectively. While this is true in general, there are special situations, for example at very low temperatures or in low dimensions where the distinction gets some what blurred leading to Bosonization or Semionic effects. Our model predicts such a Bosonization effect for Fermions, at energies corresponding to scales much larger than the Compton wavelength. Indeed such an anomalous behaviour is found experimentally in the superfluidity of He^3: Though this is sought to be explained in terms of the conventional BCS theory, the fact is that there are inexplicable anomalous features (cf.ref.[448]).

The model also predicts handedness and the blurring of Fermi-Dirac statistics in the two and one dimensional cases.

Finally it may be mentioned that very recent experimental results on carbon nanotubes [449–452] exhibit the one dimensional nature of conduction and behaviour like low temperature quantum wires thus confirming the results discussed. We make a few additional remarks.

Interestingly in the fuzzy spacetime related Quantum Mechanical Kerr-Newman Black Hole hydrodynamical vortex picture (Cf.ref.[130]) as in the usual Quantum Theory of addition of angular momenta, we can recover the fact that the sum or bound state of two such vortices or spin half particles would indeed give Bosons, so that spin half would be primary as already mentioned.

This can be seen as follows, from the theory of vortices [453]. The velocity distribution is given by

$$v = \nu/2\pi r \qquad (8.119)$$

In our case we have to use in (8.119) $v = c$ and $r = \hbar/2mc$, the Compton wavelength of the particle. So we have $\nu = \frac{h}{m}$ which is also the diffusion constant of Chapter 2.

If we consider two parallel spinning vortices separated by a distance d, then the angular velocity is given by

$$\omega = \frac{\nu}{\pi d^2}$$

whence the spin of the system turns out to be h, that is in usual units the spin is one, and the above gives the states ± 1.

There is also the case where the two above vortices are anti parallel. In this case there is no spin, but rather there is the linear velocity given by

$$v = \nu / 2\pi d$$

This corresponds to the state 0 in the spin 1 case.
Together, the two above cases give the three $-1, 0, +1$ states of spin 1 as in the Quantum Mechanical Theory.
It must be noted that the distance d between vortices could be much greater than the Compton wavelength scale, so that the wave function of the Boson in the above description would be extended in space−−it would be in the asymptotic region in comparison to the Fermionic wave function, as pointed out a little earlier. We can understand two puzzling features noted earlier, also from this point of view. The first is the fact, noticed by Darwin, as noted, that source free Electrodynamics is mathematically equivalent to the massless, force free Dirac theory. The second is the bosonic behaviour of the neutrino alluded to. In both these cases, the divide between the tight vortices, and the approximate stream lines, gets blurred. In any case, all this vindicates comments made at the beginning of this Section.
It must also be observed that in our hydrodynamical vortex picture above, when we have a bound state of the vortices, there is really no interaction in the sense of Particle Physics. Indeed the interaction description comes up once we identify the background Zero Point Field with the hydrodynamical flow. This was also seen in earlier Chapters, where interactions arose out of the background ZPF.
Finally, we now consider the magnetism of stars and planets, in the above context. It is known that Neutron stars or Pulsars have strong magnetic fields of $\sim 10^8$ Tesla in their vicinity, while certain White Dwarfs have magnetic fields $\sim 10^2$ Tesla. If we were to use conventional arguments that when a sun type star with a magnetic field $\sim 10^{-4}$ Tesla contracts, there is conservation of magnetic flux, then we are lead to magnetic fields for Pulsars and White Dwarfs which are a few orders of magnitude less than the required values[454].
We will now argue, that in the light of the above results that below the Fermi temperature, the degenerate electron gas obeys a semionic statistics, that is a statistics in between the Fermi-Dirac and Bose-Einstein, it is possible to deduce the correct magnetic fields for Neutron stars and White Dwarfs. Moreover this will also enable us to deduce the correct magnetic field of a planet like the earth.

We have for the energy density e, in case of sub Fermi temperatures,

$$e \propto \int_o^{p_F} \frac{p^2}{2m} d^3p \propto T_F^{2.5} \tag{8.120}$$

where p_F is the Fermi momentum and T_F is the Fermi temperature. On the other hand, it is known that [437, 108] in n dimensions we have,

$$e \propto T_F^{n+1} \tag{8.121}$$

(For the case $n = 3$, (8.121) is identical to the Stefan-Boltzmann law). Comparison of (8.121) and (8.120) shows that the assembly behaves with the fractal dimensionality 1.5.

Let us now consider an assembly of N electrons. As is known, if N_+ is the average number of particles with spin up, the magnetisation per unit volume is given by

$$M = \frac{\mu(2N_+ - N)}{V} \tag{8.122}$$

where μ is the electron magnetic moment. At low temperatures, in the usual theory, $N_+ \approx \frac{N}{2}$, so that the magnetisation given in (8.122) is very small. On the other hand, for Bose-Einstein statistics we would have, $N_+ \approx N$. With the above semionic statistics we have,

$$N_+ = \beta N, \frac{1}{2} < \beta < 1, \tag{8.123}$$

If N is very large, this makes an enormous difference in (8.122).

Let us first use (8.122) and (8.123) for the case of Neutron stars. In this case, as is well known, we have an assembly of degenerate electrons at temperatures $\sim 10^7 K$, (cf.for example [125]). So our earlier considerations apply. In the case of a Neutron star we know that the number density of the degenerate electrons, $n \sim 10^{31}$ per c.c.[44, 229]. So using (8.122) and (8.123) and remembering that $\mu \approx 10^{-20} G$, the magnetic field near the Pulsar is $\sim 10^{11} G <_\sim 10^8$ Tesla, as required.

Some White Dwarfs also have magnetic fields. If the White Dwarf has an interior of the dimensions of a Neutron star, with a similar magnetic field, then remembering that the radius of a White Dwarf is about 10^3 times that of a Neutron star, its magnetic field would be 10^{-6} times that of the neutron star, which is known to be the case.

Such considerations have been used by the author for the earth's magnetic field too [455].

Bibliography

[1] Sidharth, B.G. (1999). *The Celestial Key to the Vedas* (Inner Traditions, New York).

[2] Nandalas Sinha (trans.) (1986). *Vaiseshika Sutras of Kanada* (S.N. Publications, Delhi).

[3] Pannekoek, A. (1961) *A History of Astronomy* (Dover Publiations, Inc., New York).

[4] Prigogine, I. (1985). *Order Out of Chaos* (Flamingo, Harper Collins, London), pp.xix.

[5] Nicolis, G. and Prigogine, I. (1989). *Exploring Complexity* (W.H. Freeman, New York), p.10.

[6] Rohrlich, F. (1965). *Classical Charged Particles* (Addison-Wesley, Reading, Mass.), pp.145ff.

[7] Barut, A.O. (1964). *Electrodynamics and Classical Theory of Fields and Particles* (Dover Publications, New York), p.97ff.

[8] Jimenez, J.L. and Campos, I. (1999). *Found. of Phys.Lett.* **12**, 2, pp.127–146.

[9] Sidharth, B.G. *The Lorentz-Dirac and Dirac Equations* to appear in *Electromagnetic Phenomena.*

[10] Hoyle, F. and Narlikar, J.V. (1996). *Lectures on Cosmology and Action at a Distance Electrodynamics* (World Scientific, Singapore).

[11] Hoyle, F. and Narlikar, J.V. (1974). *Action at a Distance in Physics and Cosmology* (W.H. Freeman, New York), pp.12-18.

[12] Sidharth, B.G. *The Feynman-Wheeler Perfect Absorber Theory in a New Light* to appear in *Foundation of Physics.*

[13] Wheeler, J.A. and Feynman, R.P. (1945) *Rev.Mod.Phys* 17, p.157.

[14] Sidharth, B.G. (1999). *Instantaneous Action at a Distance in Modern Physics: "Pro and Contra"* A.E. Chubykalo et al. (eds.) (Nova Science Publishing, New York).

[15] Dirac, P.A.M. (1958). *The Principles of Quantum Mechanics* (Clarendon Press, Oxford), pp.4ff, pp.253ff.

[16] Sidharth, B.G. (2005). *The Universe of Fluctuations* (Springer, Netherlands).

[17] Weinberg, S. (1972). *Gravitation and Cosmology* (John Wiley & Sons, New York), p.61ff.
[18] Sidharth, B.G. (2006). *Found.Phys.Lett.* **19**, 1, pp.87ff.
[19] Sidharth, B.G. (2006) *Found.Phys.Lett.* **19**, 4, pp.399–402.
[20] Sidharth, B.G. (2006). *Found.Phys.Lett.* **19**, 6, pp.499–500.
[21] Mersini-Houghton, L. (2006). *Mod.Phys.Lett.A.* **Vol.21**, No.1, pp.1–21.
[22] Sidharth, B.G. *Physics/0608222.*
[23] Rohrlich, F. (1997). *Am.J.Phys.* **65**, (11), pp.1051–1056.
[24] Singh, V. (1988). *Schrodinger Centenary Surveys in Physics*, Singh, V. and Lal, S. (eds.) (World Scientific, Singapore).
[25] Schilpp, P.A. (1940). *Albert Einstein: Philosopher Scientist* (Library of Living Philosophers, Evanston).
[26] Weinberg, S. (1995). *The Quantum Theory of Fields, Vol I* (Cambridge University Press, Cambridge).
[27] Weinberg, S. (1996). *The Quantum Theory of Fields, Vol II* (Cambridge University Press, Cambridge).
[28] Lee, B. and Abers, W. (1978). *Gauge Theory and Neutrino Physics* Jacob, M. (ed.) (North Holland, Amsterdam).
[29] Taylor, J.C. (1978). *Gauge Theories of Weak Interactions* (Cambridge University Press, Cambridge).
[30] Lee, T.D. (1981). *Particle Physics and Introduction to Field Theory* (Harwood Academic), pp.391ff.
[31] Hooft, G.'t. (1989). *Gauge Theories of the Forces between Elementary Particles* in *Particle Physics in the Cosmos*, Richard A Carrigan. and Peter Trower, W. (eds.) *Readings from Scientific American Magazine* (W.H. Freeman and Company, New York).
[32] Regge, T. and Alfaro, V. de. (1965). *Potential Scattering* (North Holland Publishing Co., Amsterdam).
[33] Feynman, R.P. (1985). *QED* (Penguin Books, London).
[34] Martin, B.R. and Shaw, G. (1992). *Particle Physics* (John Wiley & Sons, New York).
[35] Salam, A. (1990). *Unification of Fundamental Forces* (Cambridge University Press, Cambridge). This book gives a delightful account of steps leading to electroweak unification.
[36] Pal, P.B. and Lahiri, A. (2001). *A First Book of Quantum Field Theory* (Narosa Publishing House, New Delhi).
[37] Sachs, M. (1993). *Quantum Mechanics from General Relativity: A Paradigm Shift* in *Directions in Microphysics, Fondation Louis de Broglie*, p.393ff.
[38] Weyl, H. (1962). *Space-Time Matter* (Denver Publications Inc.,New York), 282ff.
[39] Bergmann, P.G. (1969). *Introduction to the Theory of Relativity* (Prentice-Hall, New Delhi), p248ff.
[40] Moriyasu, K. (1983). *An Elementary Primer for Gauge Theory* (World Scientific, Singapore).
[41] G 't Hooft. (2008). in *A Century of Ideas* Sidharth, B.G. (ed.) (Springer,

Dordrecht).

[42] Sidharth, B.G. (1999). *Proceedings of the First International Symposium, "Frontiers of Fundamental Physics* Sidharth, B.G. and Burinskii, A. (eds.) (Universities Press, Hyderabad).

[43] Narlikar, J.V. (1993). *Introduction to Cosmology* (Cambridge University Press, Cambridge), p.57.

[44] Ruffini, R. and Zang, L.Z. (1983). *Basic Concepts in Relativistic Astrophysics* (World Scientific, Singapore), p.111ff.

[45] Misner, C.W., Thorne, K.S. and Wheeler, J.A. (1973). *Gravitation* (W.H. Freeman, San Francisco), pp.819ff.

[46] Zee, A. (1982). *Unity of Forces in the Universe Vol.II* (World Scientific, Singapore), p.40ff.

[47] Linde, A. (1983). *Phys.Lett.* 129B, pp.177–181.

[48] Coughlin, G.D. and Dodd, J.E. (1991). *The Ideas of Particle Physics* (University Press, Cambridge).

[49] Roman, P. (1965). *Advanced Quantum Theory* (Addison-Wesley, Reading, Mass.), p.31.

[50] Bogdan, P. and Rith, K. (1993). *Particles and Nuclei: An Introduction to the Physical Concepts* (Springer-Verlag, Berlin).

[51] Dirac, P.A.M. (1962). *Proc.Roy.Soc., London* A268, p.57.

[52] Veneziano, G. (1998). *The Geometric Universe* Huggett,S.A. et al. (eds.) (Oxford University Press, Oxford), p.235ff.

[53] Veneziano, G. (1974). *Physics Reports*, **9**, No.4, p.199–242.

[54] Sidharth, B.G. (2001). *Fuzzy, non commutative spacetime: A new paradigm for a new century* in *Proceedings of Fourth International Symposium on "Frontiers of Fundamental Physics"* (Kluwer Academic/Plenum Publishers, New York), p.97–108.

[55] Fogleman, G. (1987). *Am.J.Phys.* 55, (4), pp.330–6.

[56] Schwarz, J. (1982). *Physics Reports* 89, (227).

[57] Jacob, M. (1974). *Physics Reports, Reprint Volume* (North-Holland, Amsterdam).

[58] Ramond, P. (1971). *Phys.Rev.D.* 3, (10), pp.2415–2418.

[59] Kaluza, Th. (1984). *An Introduction to Kaluza-Klein Theories* (World Scientific, Singapore).

[60] Madore, J. (1995). *An Introduction to Non-Commutative Differential Geometry* (Cambridge University Press, Cambridge).

[61] Sidharth, B.G. (2005). *Foundation of Physics* vol.36, No.8, pp.1291–1294.

[62] Hooft, G't. (1997). *In Search of the Ultimate Building Bloocks* (Cambridge University Press, Cambridge).

[63] Kuhn, T. (1970). *The Structure of Scientific Revolution*, 2nd. ed. (Chicago University Press, Chicago).

[64] Wheeler, J.A. (1983). *Am.J.Phys.* 51, (5), pp.398–406.

[65] Prigogine, I. (2007). *A Century of Ideas* Sidharth, B.G. (ed.) (Springer, Netherlands).

[66] Prugovecki, E. (1984). *Found.of Phys.* Vol.14, No.12, pp.1147–1161.

[67] Babublitz, M Jr. (1997). *The Present Status of the Quantum Theory of*

Light Jeffers, S. et al. (eds.), (Kluwer Academic Publishers, Netherlands), pp.193–203.
[68] Hushwater, V. (1997). *Am.J.Phys.* 65, (5), pp.381–384.
[69] Nelson, E. (1966). *Physical Review* Vol.150, No.4, pp.1079–1085.
[70] Zastawniak, T. (1990). *Europhys.Lett.* 13, (1), pp.13–17.
[71] Serva, M. (1988). *Annals de'Institute Henri Poincare-Physique theorique* Vol.49, No.4, pp.415–432.
[72] Bacciagaluppi, G. (1999). *Nelsonian Mechanics Revisited* (Plenum Publishing Corporation, New York).
[73] Shirai, H. (1998). *Found.of Phys.* Vol.28, No.11, pp.1633–1663.
[74] Namsrai, K. (1986). *Nonlocal Quantum Field Theory and Stochastic Quantum Mechanics* (D.Reidel Publishing Company, Boston), pp.7ff.
[75] Gueret, P. and Vigier, J.P. (1982). *Found.of Phys.* Vol.12, (11), pp.1057ff.
[76] Coveney, P.V. (1988). *Nature* Vol.333, pp.409–415.
[77] Hakim, R. (1968). *J.Math.Phys.* Vol.9, No.11, pp.1805–1818.
[78] Prugovecki, E. (1995). *Quantization, Coherent States and Complex Structures* Antoine, J.P. et al. (eds.) (Plenum Press, New York).
[79] Hakim, R. (1967). *J.Math.Phys.* Vol.8, No.6, pp.1315ff.
[80] de la Pena, L. and Cetto, A.M. (1982). *Found.of Phys.* Vol.12, No.10, pp.1017–1037.
[81] De La Pena, L. and Auerbach. (1969). *J.Math.Phys.* Vol.10, No.9, pp.1620–1630.
[82] Santos, E. (1985). *Stochastic Electrodynamics and the Bell Inequalities* in *"Open Questions in Quantum Physics"* Tarozzi, G. and van der Merwe, A. (eds.) (D. Reidel Publishing Company), pp.283–296.
[83] Nelson, E. (1973). *J.of Functional Analysis* 12, pp.97–112.
[84] Kyprianidis, A. (1992). *Found.of Phys.* pp.1449-1483.
[85] Nelson, E. (1966). *Connection Between Brownian Motion and Quantum Mechanics, Lecture Notes on Physics, Vol.100* (Springer, Berlin), pp.168–179.
[86] Ord, G.N. (1996). *Int.J.Th.Phys.* Vol.35, No.2, pp.263–266.
[87] Sornette, D. (1990). *Euro J. Phys.* 11, pp.334–337.
[88] Winterberg, F. (1995). *Int.J.Mod.Phys.* Vol.34, No.10, pp.2145–2164.
[89] Guiasu, S. (1992). *Int.J.Th.Phys.* Vol.31, No.7, p.1153ff.
[90] Smolin, L. (1986). *Quantum Concepts in Space and Time*, Penrose, R. and Isham, C.J. (eds.) (OUP, Oxford), pp.147–181.
[91] Prugovecki, E. (1995). *Principles of Quantum General Relativity* (World Singapore, Singapore).
[92] De Pena, L. (1983). *Stochastic Processes applied to Physics...* Gomez, B. (ed.) (World Scientific, Singapore).
[93] Peat, F.D. (1988). *Super Strings* (Abacus, Chicago), p.21.
[94] Heisenberg, W. (1957). *Rev.Mod.Phys.* Vol.29, No.3, pp.269–278,
[95] Moller, C. (1952). *The Theory of Relativity* (Clarendon Press, Oxford) pp.170ff.
[96] Caldirola, P. (1956). *Supplemento Al Volume III, Serie X, Del Nuovo Cimento* No.2, p.297ff.

[97] Raju, C.K. (1981). *Int.J.Th.Phys.* Vol.20, No.9, pp.681–691.

[98] Sastry, R.R. *Quantum Mechanics of Extended Objects, xxx.lanl.gov/ quant-ph/9903025.*

[99] Sternglass, E.J. (1997). *The Present Status of the Quantum Theory of Light* Jeffers, S. et al. (eds.) (Kluwer Academic Publishers, Netherlands), pp.459–469.

[100] Roychowdhury, R.K. and Roy, S. (1987). *Phys.Lett.A.* Vol.123, No.9, pp.429–432.

[101] Wheeler, J.A. (1968). *Superspace and the Nature of Quantum Geometrodynamics, Battelles Rencontres, Lectures,* De Witt, B.S. and Wheeler, J.A. (eds.) (Benjamin, New York).

[102] Duff, M.J. (1998). *Scientific American* February 1998, pp.54–59.

[103] Scherk, J. (1975). *Rev.Mod.Phys.* Vol.47, No.1, January 1975, pp.123–164.

[104] Hooft, G.'t. (1996). *Classical and Quantum Gravity* 13, pp.1023–1039.

[105] Sidharth, B.G. (2000). *Chaos, Solitons and Fractals* 11, pp.1037–1039.

[106] Sidharth, B.G. (2000). *Chaos, Solitons and Fractals* 11, pp.1171–1174.

[107] Sidharth, B.G. (2000). *Chaos, Solitons and Fractals* 11, pp.1269–1278.

[108] Reif, F. (1965). *Fundamentals of Statistical and Thermal Physics* (McGraw-Hill Book Co., Singapore).

[109] Armour, R.S. Jr. and Wheeler, J.A. (1983). *Am.J.Phys.* 51, (5), pp.405–406.

[110] Singh, J. (1961). *Great Ideas and Theories of Modern Cosmology* (Dover, New York), pp.168ff.

[111] Abbott, L.F. and Wise, M.B. (1981). *Am.J.Phys.* 49, pp.37-39.

[112] Mandelbrot, B.B. (1982). *The Fractal Geometry of Nature* (W.H. Freeman, New York), pp.2,18,27.

[113] Allen, A.D. (1983). *Speculations in Science and Technology* Vol.6, No.2, pp.165–170.

[114] Ord, G.N. (1997). *The Present Status of the Quantum Theory of Light,* Jeffers, S. et al. (eds.) (Kluwer Academic Publishers, Netherlands), pp.169–180.

[115] Ord, G.N. (1997). *The Present Status of the Quantum Theory of Light,* Jeffers, S. et al. (eds.) (Kluwer Academic Publishers, Netherlands), pp.165–168.

[116] Ord, G.N. (1983). *J.Phys.A:Math.Gen.* 16, pp.1869–1884.

[117] Ord, G.N. (1992). *Int.J.Th.Phys.* Vol.31, No.7, pp.1177–1195.

[118] Ord, G.N. (1998). *04817 Elsevier Science CHAOS* Ms 1036, MFC September 1998 (Chaos, Solitons and Fractals).

[119] Ord, G.N. and Gualtieri, J.A. (1998). *02651 ElSevier Chaos* Ms 870 May 1998 (Chaos, Solitons and Fractals).

[120] Nottale, L. (1995). *Quantum Mechanics, Diffusion and Chaotic Fractals* El Naschie, M.S., Rossler, O. and Prigogine, I. (eds.) (ElSevier, Oxford), pp.51–78.

[121] Nottale, L. (1994). *Chaos, Solitons & Fractals* 4, 3, pp.361–388 and references therein.

[122] Feynman, R.P. and Hibbs, A.R. (1965). *Quantum Mechanics and Path*

Integrals (McGraw-Hill).
[123] Ginzburg, V.L. (1976). *Key Problems of Physics and Astrophysics* (Mir Publishers, Moscow).
[124] Einstein, A. (1999). *J. Franklin Inst., quoted by M.S. El Naschie, Chaos Solitons and Fractals* 10, (2-3), p.163.
[125] Huang, K. (1975). *Statistical Mechanics* (Wiley Eastern, New Delhi), pp.75ff.
[126] Hayakawa, S. (1965). *Suppl of PTP Commemmorative Issue* pp.532-541.
[127] Sidharth, B.G. (2000). *Chaos, Solitons and Fractals* (12), (1), pp.173–178.
[128] Lucas, J.R. (1984). *Space Time, And Causality* (Oxford Clarendon Press).
[129] Sidharth, B.G. (1997). *Ind.J. of Pure and Applied Physics* 35, pp.456ff.
[130] Sidharth, B.G. (2001). *Chaotic Universe: From the Planck to the Hubble Scale* (Nova Science, New York).
[131] Nottale, L. (1993). *Fractal Space-Time and Microphysics: Towards a Theory of Scale Relativity* (World Scientific, Singapore), pp.312.
[132] Joos, G. (1951). *Theoretical Physics* (Blackie, London), pp199ff.
[133] Gibson, J.G. (1968). *Proc.Camb.Phil.Soc.* 64, pp.1061.
[134] Kracklauer, A.F. (1974). *Phys.Rev.D.* 10, pp.1358.
[135] Cavalleri, G. (1981). *Phys.Rev.D.* 23, APPB pp.363.
[136] Cavalleri, G. and Mauri, G. (1990). *Phys.Rev.B.* 41, 2, pp.6751,
[137] Bosi, L. and Cavelleni, G. (2002). *Nuovo Cimento B.* 117, pp.243.
[138] Balescu, R. (1975). *Equilibrium and Non Equilibrium Statistical Mechanics* (John Wiley, New York).
[139] Rueda, A. and Haisch, B. (1998). *Found. of Phys.* Vol.28, No.7, pp.1057–1108.
[140] Sivaram, C. (1983). *Am.J.Phys.* 51 (3), pp.277.
[141] Sivaram, C. (1992). *Am.J.Phys.*, 50 (2), pp.279.
[142] Ichinose, T. (1984). *Physica* 124A, pp.419.
[143] Sachs, M. (1982). *General Relativity and Matter* (D. Reidel Publishing Company, Holland), pp.45ff.
[144] Powell, J.L. and Crasemann, B. (1988). *Quantum Mechanics* (Narosa Publishing House, New Delhi), pp.5ff.
[145] Snyder, H.S. (1947). *Physical Review* Vol.72, No.1, July 1 1947, pp.68–71.
[146] Snyder, H.S. (1947). *Physical Review* Vol.71, No.1, January 1 1947, pp.38–41.
[147] Kadyshevskii, V.G. (1962). *Translated from Doklady Akademii Nauk SSSR* Vol.147, No.6 December 1962, pp.1336–1339.
[148] Schild, A. (1948). *Phys.Rev.* 73, pp.414–415.
[149] Dirac, P.A.M. (1978). *Directions in Physics* (John Wiley, New York).
[150] Bombelli, L., Lee, J., Meyer, D. and Sorkin, R.D. (1987). *Physical Review Letters* Vol.59, No.5, August 1987, pp.521–524.
[151] Finkelstein, D.R. (1996). *Quantum Relativity A Synthesis of the Ideas of Einstein and Heisenberg* (Springer, Berlin).
[152] Wolf, C. (1990). *Hadronic Journal* Vol.13, pp.22–29.
[153] Lee, T.D. (1983). *Physics Letters* Vol.122B, No.3,4, 10 March 1983, pp.217–220.

[154] Sidharth, B.G. (2003) *Found.Phys.Lett.* 16, (1), pp.91–97.

[155] Shirokov, Yu. M. (1958). *Soviet Physics JETP* 6, (33), No.5, pp.929–935.

[156] Heine, V. (1960). *Group Theory in Quantum Mechanics* (Pergamon Press, Oxford), pp.364.

[157] Sidharth, B.G. (2002). *Found.Phys.Lett.* 15, (6), pp.577–583.

[158] Sidharth, B.G. (2004). *Found.Phys.Lett.* 17, (5), pp.503–506.

[159] Penrose, R. (1971). *Angular Momentum: An approach to combinational space-time* in *Quantum Theory and Beyond*, Bastin, T. (ed.) (Cambridge University press, Cambridge), pp.151ff.

[160] Newman, E.T. (1973). *J.Math.Phys* 14, (1), pp.102.

[161] Newman, E.T. (1975). *Proceedings of Enrico Fermi International School of Physics*, pp.557.

[162] Zakrewski, S. (1995). *Quantization, Cohrent States and Complex Structures* Antoine, J.P. et al, (eds.) (Plenum Press, New York), pp.249ff.

[163] Sidharth, B.G. (2003). *An Interface between Classical Electrodynamics and Quantum Mechanics*, in *Has the last word been said on Classical Electrodynamics?* Chubykalo, A. et al. (eds.) (Rinton Press, USA).

[164] Nottale, L. (1996). *Chaos, Solitons and Fractals* 7, (6), pp.877ff.

[165] Itzykson, C. and Zuber, J. (1980). *Quantum-Field Theory* (Mc-Graw Hill, New York), pp.139.

[166] Sidharth, B.G. (2002). *Chaos, Solitons and Fractals* 13, pp.189–193.

[167] Tumulka, R. (2006). *Eur.J.Phys.* 26, pp.l11-l13.

[168] Pauli, W. (1980). *General Principles of Quantum Mechanics*, trans. Achutham, P. and Venkatesam, K. (Springer-Verlag, New York), pp.63.

[169] Park, D. (1986). *Fundamental Questions in Quantum Mechanics* Roth, L.M. and Inomata, A. (eds.) (Gordon & Breach Science Publishers, New York), pp.263ff.

[170] Sidharth, B.G. (2007). *Encounters:Abdus Salam* in *New Advances in Physics* Vol.1, No.1, March 2007, pp.1–17.

[171] Science, December 2003

[172] Linde, A.D. (1982). *Phys.Lett.* 108B, 389.

[173] Sidharth, B.G. (1998). *Frontiers of Quantum Physics (1997)* Lim, S.C., et al. (eds.) (Springer Verlag, Singapore).

[174] Sidharth, B.G. (1999). *Proc. of the Eighth Marcell Grossmann Meeting on General Relativity (1997)* Piran, T. (ed.) (World Scientific, Singapore), pp.476–479.

[175] Sidharth, B.G. (1998). *Int.J. of Mod.Phys.A* 13, (15), pp.2599ff.

[176] Sidharth, B.G. (1998). *International Journal of Theoretical Physics* Vol.37, No.4, pp.1307–1312.

[177] Perlmutter, S., et al. (1998). *Nature* Vol.391, 1 January 1998, pp.51–59.

[178] Kirshner, R.P. (1999). *Proc. Natl. Acad. Sci. USA* Vol.96, April 1999, pp.4224–4227.

[179] Musser, G. and Alpert, M. (2000). *Scientific American* January 2000, pp.27.

[180] Caldwell, R.R. and Steinhardt, P.J. (2000). *Physics World* November 2000, pp.31.

[181] Bjorken, J.D. and Drell, S.D. (1964). *Relativistic Quantum Mechanics* (McGraw Hill, New York), pp.39.
[182] Sidharth, B.G. (2006). in *Einstein and Poincare* Dvaeglazov, V. et al. (eds.) (Apeiron Press, Canada).
[183] Wilczek, F. (1999). *Physics Today* January 1999, pp.11.
[184] Milonni, P.W. (1994). *The Quantum Vacuum: An Introduction to Quantum Electrodynamics* (Academic Press, San Diego).
[185] Cole, D.C. (1983). *Essays on the Formal Aspects of Electromagnetic Theory* Lakhtakia, A. (ed.) (World Scientific, Singapore), pp.501ff.
[186] Podolny, R. (1983). *Something Called Nothing* (Mir Publishers, Moscow).
[187] Zeldovich, Ya. B. (1967). *JETP Lett.* 6, pp.316.
[188] Weinberg, S. (1979). *Phys.Rev.Lett.* 43, pp.1566.
[189] Mostepanenko, V.M. and Trunov, N.N. (1988). *Sov.Phys.Usp.* 31, (11), November 1988, pp.965–987.
[190] Lamoreauz, S.K. (1997). *Phys.Rev.Lett.* Vol.78, No.1, January 1997, pp.5–8.
[191] Petroni, N.C. and Vigier, J.P. (1983). *Foundations of Physics* Vol.13, No.2, pp.253–286.
[192] Raiford, M.T. (1999). *Physics Today* July 1999, pp.81.
[193] Milonni, P.W. (1988). *Physica Scripta.* Vol. T21, pp.102–109.
[194] Milonni, P.W. and Shih, M.L. (1991). *Am.J.Phys.* 59, (8), pp.684–698.
[195] Lee, T.D. (1981). *Statistical Mechanics of Quarks and Hadrons* H. Satz, H. (ed.) (North-Holland Publishing Company), pp.3ff.
[196] Achuthan, P. et al. (1980). *Gravitation, Quanta and the Universe* Prasanna, A.R., Narlikar, J.V. and Vishveshwara, C.V. (eds.) (Wiley Eastern, New Delhi), pp.300.
[197] Sidharth, B.G. (2002). *Chaos, Solitons and Fractals* 14, pp.167–169.
[198] Sidharth, B.G. (2003). *Chaos, Solitons and Fractals* 15, (1), pp.25–28.
[199] Castell, L. (1980). *Quantum Theory and Gravitation* Marlow, A.R. (ed.) (Academic Press, New York), pp.147ff.
[200] Sidharth, B.G. (2003). *Chaos, Solitons and Fractals* 16, (4), pp.613–620.
[201] Melnikov, V.N. (1994). *Int.J.of Th.Phys.* 33, (7), pp.1569–1579.
[202] Science, December 1998
[203] Barrow, J.D. and Parsons, P. (1997). *Phys.Rev.D.* Vol.55, No.4, 15 February 1997, pp.1906ff.
[204] Narlikar, J.V. (1983). *Foundations of Physics* Vol.13, No.3, pp.311–323.
[205] Narlikar, J.V. (1989). *Did the Universe Originate in a Big Bang?* in *Cosmic Perspectives* Biswas, S.K., Mallik, D.C.V. and Vishveshwara, C.V. (eds.) (Cambrdige University Press, Cambridge), pp.109ff.
[206] Sidharth, B.G. (2008). *Ether, Space-Time and Gravity* **Vol.3** Michael Duffy (ed.) (Hadronic Press, USA) (in press).
[207] Sidharth, B.G. (2003). *Chaos, Solitons and Fractals* 18, (1), pp.197–201.
[208] Sidharth, B.G. (2000). *Nuovo Cimento* **115B** (12), (2), pp.151ff.
[209] Goldstein, H. (1966). *Classical Mechanics* (Addison-Wesley, Reading, Mass.), pp.76ff.
[210] Kuhne, R.W. *xxx.lanl.gov/gr-qc/9809075.*

[211] Berger, A.L. (1976). *Astronomy and Astrophysics* 51, pp.127–135.
[212] Anderson, J.L. et al. *xxx.lanl.gov/gr-qc/9808081.*
[213] Denman, H.H. (1983). *Am.J.Phys.* 51, (1), pp.71.
[214] Silverman, M.P. (1980). *Am.J.Phys.* 48, pp.72.
[215] Brill, D.R. and Goel, D. (1999) *Am.J.Phys.* 67, (4), pp.317.
[216] Lass, H. (1950). *Vector and Tensor Analysis* (McGraw-Hill Book Co., Tokyo), pp.295 ff.
[217] Sivaram, C. and Sabbata, V. de. (1993). *Foundations of Physics Letters* 6, (6).
[218] Sidharth, B.G. (2001). *Chaos, Solitons and Fractals* 12, pp.1101–1104.
[219] Sidharth, B.G. and Popova, A.D. (1996). *Differential Equations and Dynamical Systems* (DEDS), 4, (3/4), pp431–440.
[220] Milgrom, M. (2001). *MOND - A Pedagogical Review, Presented at the XXV International School of Theoretical Physics "Particles and Astrophysics-Standard Models and Beyond"* (Ustron, Poland).
[221] Milgrom, M. (1983). *APJ* 270, pp.371.
[222] Milgrom, M. (1986). *APJ* 302, pp.617.
[223] Milgrom, M. (1989). *Comm. Astrophys.* 13:4, pp.215.
[224] Milgrom, M. (1994). *Ann.Phys.* 229, pp.384.
[225] Milgrom, M. (1997). *Phys.Rev.E* 56, pp.1148.
[226] Anderson, J.D. et al. (2002). *Phys.Rev.D* 65, pp.082004ff.
[227] Sidharth, B.G. (2006). *Found.of Phys.Lett.* 19, (6), pp.611–617.
[228] Davies, P. (1989). *The New Physics* Davies, P. (ed.) (Cambridge University Press, Cambridge), pp.446ff.- New Physics
[229] Ohanian, C.H. and Ruffini, R. (1994). *Gravitation and Spacetime* (New York, 1994), pp.397.
[230] Uzan, J.P. (2003). *Rev.Mod.Phys.* 75, April 2003, pp.403–455.
[231] Narlikar, J.V. (1977). *The Structure of the Universe* (Oxford University Press, Oxford), pp.175.
[232] Bennett, C.L. et al. (2003). *Astrophys. J. Suppl.* 148.
[233] Hinshaw, G. et al. (2003). *Astrophys. J. Suppl.* 148, pp.135.
[234] Sidharth, B.G. (2006). *Int.J.Mod.Phys.E.* 15, (1), pp.255ff.
[235] McGauge, S.S. (1999). *APJ Lett.* 523, pp.L99.
[236] H. Kragh. (1990). *Dirac A Scientific Biography* (Cambridge University Press, New York), pp.223ff. Also Dirac, P.A.M. (1978). *Directions in Physics* Hora, H. and Shepanski, J.R. (eds.) (John Wiley & Sons, New York), pp.76–77.
[237] Webb, J.K. et al. (2001). *Phys.Rev.Lett.* 87, (9), pp.091301–1 ff.
[238] Webb, J.K. et al. (1999) *Phys.Rev.Lett.* 82, pp884ff.
[239] Kuhne, R.W. (1999). *Mod.Phys.Lett.A.* Vol.14, No.27, pp.1917–1922.
[240] Kalyana Rama, S. (2001). *Phys.Lett. B* 519, pp.103–110.
[241] Moffat, J.W. (1993). *Int.J.Mod.Phys.D* 2, pp.351; Moffat, J.W. (1993). *Found.Phys.23.*
[242] Magueijo, J. (2000). *Phys.Rev.D.* 62, pp.103521.
[243] Witten, E. (1996). *Physics Today* April 1996, pp.24–30.
[244] Ne'eman, Y. (1999). *Proceedings of the First Internatioinal Symposium,*

Frontiers of Fundmental Physics, Sidharth, B.G. and Burinskii, A. (eds.) (Universities Press, Hyderabad), pp.83ff.
[245] Sidharth, B.G. (2001). *Chaos, Solitons and Fractals* 12, pp.1449–1457.
[246] Goodstein, D.L. (1975). *States of Matter* (Dover Publications, Inc., New York), pp.462ff.
[247] Sidharth, B.G. (1994). *Nonlinear World*1, pp.403–408.
[248] Sidharth, B.G. (1999). *J.Stat.Phys.* 95, (3/4), pp.775–784.
[249] Georgi, H. (1989). *The New Physics* Davies, P. (ed.) (Cambridge University Press, Cambridge), pp.446ff.
[250] Wilson, K.G. (1983). *Rev.Mod.Phys.* 55, pp.583ff.
[251] Sidharth, B.G. (2004). *Chaos, Solitons and Fractals* 20, pp.701–703.
[252] Baez, J. (2003). *Nature* Vol.421, February 2003, pp.702–703.
[253] Sivaram, C. (1982). *Astrophysics and Space Science* 88, pp.507–510.
[254] Landsberg, P.T. (1983). *Am.J.Phys.* 51, pp.274–275.
[255] Sachidanandam, S. (1983). *Physics Letters* Vol.97A, No.8, 19 September 1983, pp.323–324.
[256] Boyer, T.H. (1968). *Phys.Rev.* 174, pp.1631.
[257] Rosen, N. (1993). *Int.J.Th.Phys.*, 32, (8), pp.1435–1440.
[258] Kolb, E.W. and Pecci, R.D. (1994). *Proceedings of the 1994 Snowmass Summer Study "Particle and Nuclear Astrophysics and Cosmology in the Next Millennium"* (World Scientific, Singapore).
[259] Zang, L.Z. and Ruffini, R. (1984). *Cosmology of the Early Universe* (World Scientific, Singapore, 1984), pp.93.
[260] Trotta, R. and Melchiorri, A. *www-astro.physics.ox.ac.uk/ rxt/pdf/CNB.*
[261] Weiler, T.J. (2005). *Int.J.Mod.Phys.A* **20**, (6), pp.1168–1179.
[262] Goldsmith, D. (1995). *Einstein's Greatest Blunder: The Cosmological Constant and other Fudge Factors in the Physics of the Universe* (Harvard University Press, Harvard,1995).
[263] Perlmutter, S. et al. *astro-ph/981213.*
[264] Carroll, S.M. (2001). *astro-ph/0004075 Living.Rev.Rel* **4**.
[265] Perlmutter, S., Turner, M.S. and White, M. (1999). *Phys.Rev.Lett* **83**, pp.670.
[266] Inaba, M. (2001). *Int.J.Mod.Phys.A.* 16, (17), pp.2965–73.
[267] Elnaschie, M.S. (1998). *A note on quantum mechanics, diffusional intereference and informions. Chaos, Solitons and Fractals* 5, (5), pp.881–4.
[268] Santamato, E. (1984). *Phys.Rev.D.* 29, (2), pp.216ff.
[269] Santamato, E. (1984). *J.Math.Phys.* 25, (8), pp.2477ff.
[270] Santamato, E. (1985). *Phys.Rev.D.* 32, (10), pp.2615ff.
[271] Santamato, E. (1988). *Phys.Lett.A.*130, (4 & 5), 199ff.
[272] Sidharth, B.G. (2006). *Int.J.SCI.* January 2006, pp.42ff.
[273] Jack Ng, Y. and Van Dam, H. (1994) *Mod.Phys.Lett.A.* 9, (4), pp.335–340.
[274] Sidharth, B.G. (Ed.) (2008). *A Century of Ideas* (Springer, Dordrecht).
[275] Cercignani, C. (1998). *Found.Phys.Lett.* Vol.11, No.2, pp.189-199.
[276] Cercignani, C., Galgani, L. and Scotti, A. (1972). *Phys.Lett.* 38A, pp.403.
[277] Sidharth, B.G. (2005). *Found.Phys.Lett.* 18, (4), pp.387–391.
[278] Sakharov, A.D. (1968). *Soviet Physics - Doklady* Vol.12, No.11, pp.1040–

1041.
[279] Golfain, E. (2004). *Chaos, Solitons and Fractals* 22, (3), pp.513-520.
[280] De Broglie, L. and Vigier, J.P. (1972). *Phys.Rev.Lett.* 28, pp.1001–1004.
[281] De Broglie. (1940). *La mecanique ondulatoire du photon Une nouvelle theorie de la lumiere* Vol.I (Paris, Hermann).
[282] De Broglie. (1942). *Les interactions entre les photons et la matiere* Vol.II (Paris, Hermann).
[283] Deser, S. (1972). *Ann Inst. Henri Poincare, Vol.XVI* (Paris, Gauthier-Villors), pp.79.
[284] Lakes, R. (1998). *Phys.Rev.Lett.* 80, (9), pp.1826ff.
[285] Marshall, T.W. (1997). *New Developments on Fundamental Problems in Quantum Physics* Ferrero, M. and Van der Merwe, A. (eds.) (Kluwer, Dordrecht), pp.231ff.
[286] Marshall, T.W. and Santos, E. (1997). *The Present Status of the Quantum Theory of Light*, Jeffers, S. et al. (eds.) (Kluwer Academic, Dordrecht), pp.67–77.
[287] Armstrong, H.L. (1983). *Am.J.Phys.* 51, (2), February 1983, pp.103.
[288] Sachs, M. (1997). *The Present Status of the Quantum Theory of Light* Jeffers, S. et al. (eds.), (Kluwer Academic, Dordrecht), pp.79–96.
[289] Sachidanandam, S. and Raghavacharyulu, I.V.V. (1983). *Ind.J.Pure and Appl.Phys.*, Vol.21, July 1983, pp.408–412.
[290] Pavlopoulos, T.G. (2005). *Phys.Lett.B.* 625, pp.13-18.
[291] Vigier, J.P. (1990). *IEEE Transactions of Plasma Science* 18, (1), pp.64–72.
[292] Sidharth, B.G. *physics/0607208.*
[293] Gersten, A. (1999) *Found.Phys.Lett.* 12, (3), pp.291–298.
[294] Dvoeglazov, V.V. and Gonzalez, J.L.Q. (2006). *Found.Phys.Lett.* 19, (2), pp.195ff.
[295] Ignatiev, A. Yu. and Joshi, G.C. (1996). *Mod.Phys.Lett.A.* 11, pp.2735–2741.
[296] Tucker, R.H. and Hammer, C.L. (1978). *Phys.Rev.D.* Vol.3, No.10, 2448ff.
[297] Sidharth, B.G. (2003) *Nuovo Cimento B* **118B**, (1), pp.35–40.
[298] Dirac, P.A.M. (1982). *Monopoles in Quantum Field Theory* Craigie, N.S., Goddard, P. and Nahm, W. (eds.) (World Scientific, Singapore), pp.iii.
[299] Sidharth, B.G. (2002). *Annales de la Fondation Louis de Broglie* 27, (2), pp.333ff.
[300] Sidharth, B.G. (2005). *Int.J.Mod.Phys.E.* 14, (2), p.215ff.
[301] Evans, M. and Vigier, J.P. (1995). *The Enigmatic Photon* (Kluwer Academic, Dordrecht), pp.136.
[302] Bartlete, D.F. and Corlo, T.R. (1985). *Phys.Rev.Lett.* Vol.55, pp.49.
[303] Mott, N.F. and Massey, H.S.W. (1965) *The Theory of Atomic Collisions* (Oxford University Press, Oxford), pp.53-68.
[304] Joachain, C.J. (1975). *Quantum Collision Theory* (North-Holland Publishing Co., Amsterdam), pp.133–146.
[305] Sidharth, B.G. and Abdel-Hafez, A. (1979). *Acta Phys.Pol.* **A 56**, pp.577.
[306] Sidharth, B.G. and Abdel-Hafez, A. (1980) *Acta Phys.Pol.* **A 57**, pp.287ff.
[307] Sidharth, B.G. (1983). *J.Math.Phys.* **24**, pp.878.

[308] Greiner, W., Muller, B. and Rafelski, J. (1985). *Quantum Electrodynamics of Strong Fields* (Springer-Verlag, Berlin).
[309] Terezawa, H. (1996). *Mod.Phys.Lett.A.* Vol.11, No.38, pp.2971–2976.
[310] Hagiwara, K. et al. (2002). *(Particle Data Group), Phys.Rev.B.* 66, 010001.
[311] Sidharth, B.G. (1999). *Mod.Phys.Lett.A.* Vol.14, No.5, pp.387–389.
[312] Sidharth, B.G. (1997). *Mod.Phys.Lett.A.* Vol.12, No.32, pp.2469–2471.
[313] Sidharth, B.G. and Lobanov, Y. Yu. (1998). *Frontiers of Fundamental Physics* Sidharth, B.G. and Burinskii, A. (eds.) (Universities Press), pp.68ff.
[314] Sidharth, B.G. (2005). *Hadronic Journal* 28, 5, October 2005, pp.599ff.
[315] Amati, D. (1992). *Sakharov Memorial Lectures* Kaddysh, L.V. and Feinberg, N.Y. (eds.) (Nova Science, New York), pp.455ff.
[316] Kempf, A. (2000). *From the Planck length to the Hubble radius* Zichichi, A. (ed.) (World Scientific, Singapore), pp.613ff.
[317] Madore, J. (1992) *Class.Quantum Grav.* 9, p.69-87.
[318] Sidharth, B.G. (2001). *Concise Encyclopaedia of SuperSymmetry and Non Commutative Structures in Mathematics and Physics* Bagger, J., Duply, S. and Sugel, W. (eds.) (Kluwer Academic Publishers, New York).
[319] Schiff, L.I. (1968). *Quantum Mechanics* (McGraw Hill, London).
[320] Davydov, A.S. (1965). *Quantum Mechanics* (Pergamon Press, Oxford), pp.655.
[321] Rai Choudhuri, A. (1999). *The Physics of Fluids and Plasma* (Cambridge University Press) pp.137.
[322] Batchelor, G.K. (1993). *Fluid Mechanics* (Cambridge University Press, New Delhi).
[323] Vasudevan, R. (1994). *Hydrodynamical Formulation of Quantum Mechanics*, in *Perspectives in Theoretical Nuclear Physics*, Srinivas Rao, K., and Satpathy L. (eds.) (Wiley Eastern, New Delhi), pp.216ff.
[324] Sidharth, B.G. (2001). *Nuovo Cimento* **116B**, (6), pp.4ff.
[325] Simmons, G.F. (1965). *Introduction to Topology and Modern Analysis* (McGraw Hill Book Co. Inc., New York), pp.135.
[326] Altaisky, M.V. and Sidharth, B.G. (1999). *Chaos, Solitons and Fractals* Vol. 10, No.2-3, pp.167-176.
[327] Munkres, J.R. (1988). *Topology*(Prentice Hall India, New Delhi).
[328] Gullick, D. (1997). *Encounters With Chaos* (McGraw Hill, New York), p.114ff.
[329] Sidharth, B.G. (1998). *Gravitation and Cosmology* 4, (2), (14), pp.158ff.
[330] Gleick, J. (1987). *Chaos: Making a New Science* (Viking, New York), p.321.
[331] Sidharth, B.G. (2003). *Chaos, Solitons and Fractals* 15, pp.593-595.
[332] Elnaschie, M.S. (1996). *Chaos, Solitons and Fractals* 7, (4), pp.499–518.
[333] Sidharth, B.G. (2005). *Chaos, Solitons and Fractals* 25, pp.965–968.
[334] Frisch, D.H. and Thorndike, A.M. (1963). *Elementary Particles* (Van Nostrand, Princeton), p.96ff.
[335] Sidharth, B.G. (2004). *Chaos, Solitons and Fractals* 22, pp.537-540.
[336] Ho, Vu B. and Morgan J Michael. (1996). *J.Phys.A: Math. Gen.* **29** pp.1497–1510.

[337] Sidharth, B.G. (2004). *Annales Fondation L De Broglie* 29, (3), pp.1.
[338] Sidharth, B.G. (2004). *Int.J.of Th.Phys.* Vol.43, (9), September 2004, pp.1857-1861.
[339] Sidharth, B.G. (2005). *Int.J.Mod.Phys.E.* 14, (6), pp.923ff.
[340] Magueijo, J. (2003). *Physics Reports* (66), pp.2025.
[341] Jacobson, T., Liberati, S., and Mattingly, D. *hep-ph/0407-370.*
[342] Jacobson, T., Liberati, S. and Mattingly, D. (2006) *Annales of Phys.* **32L**, pp.150-196.
[343] Amelino-Camelia, G. et al. *gr-qc/0501053* and (2005). *AIP Conference Proceedings* Vol.758, April 4, 2005, pp.30–80.
[344] Gonzales Mestres, L. *Physics/9704017.*
[345] Coleman, S. and Glashow, S.L. (1999). *Phys.Rev. D.* 59, pp.116008.
[346] Jacobson, T. *xxx.astro-ph/0212190.*
[347] Olinto, A.V. (2000). *Phys.Rev.* 333-334, pp.329ff.
[348] Carroll, S.M. (2001). *Phys.Rev.Lett.* 87, pp.141601ff.
[349] Nagano, M., Rev.Mod.Phys., 72, 2000, pp.689ff.
[350] Montvay, I. and Munster, G. (1994). *Quantum Fields on a Lattice* (Cambridge University Press) pp.174ff.
[351] Kifune, T. (1999). *astro-ph/9904164*; *Astrophys. J. Lett.* 518, pp.L21.
[352] Protheroe, R.J. and Meyer, H. (2000). *Phys.Lett.* B493, pp.1.
[353] Aloisio, R., Blasi, P., Ghia, P.L. and Grillo, A.F. (2000). *Phys.Rev.* D62, pp.053010.
[354] Kluzniak, W. *astro-ph/9905308.*
[355] Sato, H. *astro-ph/0005218.*
[356] Amelino-Camelia, G. and Piran, T. (2001). *Physics Letters* B497, pp.265–270.
[357] Amelino-Camelia, G. *gr-qc/0012051v2* (He proposes a conceptual framework in which deformed dispersion relations coexist with a relativistic description of the short distance structure of spacetime).
[358] Amelino-Camelia, G. and Piran, T. *Phys.Rev. D.* Vol.64, pp.036005.
[359] Amelino-Camelia, G., John Ellis, Mavnomatos, N.E., Nanopoulos, D.V. and Subir Sarkar. (1998). *Nature* 393, 25 June, 1998, pp.763-765; *(astro-ph/9712103 v2 17 April 1998).*
[360] Amelino-Camelia, G. (2002). *Nature* Vol.418, 4 July 2002.
[361] *http://glast.gsfc.nasa.gov/.*
[362] Sidharth, B.G. *Different Routes to Lorentz Symmetry Variations* to appear in *Foundation of Physics.*
[363] Goenner, H.F. and Bogoslovsky, Yu G. (1999). *General Relativity and Gravitation* Vol.31, No.9, pp.1383–1394.
[364] Bogoslovsky, Yu G. and Goenner, H.F. (1999). *General Relativity and Gravitation* Vol.31, No.10, 1565–1603.
[365] Bogoslovsky, Yu G. and Goenner, H.F. (1998). *Physics Letters A.* 244, pp.222–228.
[366] Don Colladay. (2005). *Int.Mod.Phys.A.* Vol.20, No.6, pp.1260–1267.
[367] Sidharth, B.G. (2006). *Electromagnetic Phenomena* Vol.6, No.1,(16), pp.63ff.

[368] Cardone, F., Mignani, R. and Scrimaglio, R. (2006). *Foundation of Physics* Vol.36, No.2, pp.263–290.

[369] Cardone, F. and Mignani, R. (2007). *Deformed Spacetime: Geometrizing Interactions in Four and Five Dimensions* (Springer, Fundamental Theories of Physics), 157.

[370] Cardone, F. and Mignani, R. (2004). *Energy and Geometry: An Introduction to Deformed Special Relativity* (World Scientific, Series in Contemporary Chemical Physics), Vol.22.

[371] Sidharth, B.G. (2002). *Nuovo Cimento* **117B**, (6), 2002, pp.703ff.

[372] Saito, T. (2000). *Gravitation and Cosmology* 6, No.22, pp.130–136.

[373] Landau, L.D. and Lifshitz, E.M. (1976). *Quantum Mechanics* (Pergamon Press, Oxford) pp.456ff.

[374] Girotti, H.O. (2004). *Am.J.Phys.* 72, (5), May 2004, pp.608–612.

[375] Dyson, F.J. (1988). *Infinite in all directions* (Harper and Row, New York).

[376] Jackson, J.D. (1975) *Classical Electrodynamics* (John Wiley, New York).

[377] Cooper, L.N. (1968). *An Introduction to the Meaning and Structure of Physics* (Harper International, New York), pp.309ff.

[378] Einstein, A. (1965). *The Meaning of Relativity* (Oxford and IBH, New Delhi), pp.93–94.

[379] Elias, V., Pati, J.C. and Salam, A. (1977). *Centre for Theoretical Physics, University of Maryland, Physics Publication*, September 1977, pp.78–101,.

[380] Weinberg, S., Scientific American, December 1999, pp.36ff.

[381] Kolb, E.W., Seckel, D. and Turner, M.S. (2985). *Nature* Vol.314, 4, April 1985, pp.415ff.

[382] Nambu, Y. (1952). *Letters to the Editor, PTP* May 14 1952, pp.595.

[383] Schwarz, J., Green, M.B. and Witten, E. (1987). *SuperString Theory* Vol.I (Cambridge University Press, Cambridge).

[384] Schwarz, J., Green, M.B. and Witten, E. (1987). *SuperString Theory* Vol.II (Cambridge University Press, Cambridge).

[385] Sidharth, B.G. (2005). *Annales Fondation L. De Broglie* 30, (2), pp.151–156.

[386] Landau, L.D. and Lifshitz, E.M. (1975). *The Classical Theory of Fields* (Pergamon Press, Oxford), pp.77ff.

[387] Smit, J. (2002). *Quantum Fields on a Lattice* (Cambridge University Press), pp.2ff.

[388] Nambu, Y. (1981). *Quarks - Frontiers in Elementary Particle Physics* (World Scientific, Singapore) pp.212.

[389] Sidharth, B.G. (2007). *The Common Origin of Gravitation, General Relativity, Electromagnetism and Quantum Theory* in *Proceedings of the Eighth International Symposium on 'Frontiers of Fundamental Physics'* (American Institute of Physics, Melville, New York), pp.195ff.

[390] Bade, W.L. and Herbert Jehle. (1952). *Reviews of Modern Physics* **25**,(3), pp.714ff.

[391] Pavsic, M. (2005). *Foundations of Physics* Vol.35, No.9, September 2005, pp.1617–1641 and references therein.

[392] Polchinski, J. (1995). *Phys.Rev.Lett.* 75, (26), pp.4724–27.

[393] Barut, A.O. and Pavsic, M. (1992). *Mod.Phys.Lett.*, pp.381.
[394] Pavsic, M. (1991). *Nuovo Cim. A* 104, pp.1337.
[395] E. Merzbacher. (1970). *Quantum Mechanics* (Wiley, New York); Greiner, W. (1983). *Relativistic Quantum Mechanics: Wave Equation* 2nd Ed. (Springer); Greiner, W. and Reinhardt, J. (1987). *Quantum Electrodynamics* 3rd Ed. (Springer).
[396] Hawking, S.W. and Werner Israel. (1987). *300 Years of Gravitation* (Cambridge University Press, Cambridge).
[397] Greene, B. (1999). *The Elegant Universe* (Vintage, London), pp.15.
[398] DeWitt, B.S. (1967). *Phy Rev.* 160, pp.1113.
[399] S.S. Schweber. (1961). *An Introduction to Relativistic Quantum Field Theory* (Harper and Row, New York).
[400] B.G. Sidharth. (2006). *Int.J.Mod.Phys.A.* 21, (31), December 2006, pp.6315.
[401] Sivaram, C. and Sinha, K.P. (1974). *Lettere Il Nuovo Cimento* Vol.10, No.6, pp.227–230.
[402] Sidharth, B.G. (2005). *Chaos, Solitons and Fractals* 24, pp.443–445.
[403] Feynman, R.P. (1995). *Feynman Lectures on Gravitation* (Addison-Wesley Publishing Company, Reading, mass.).
[404] Sidharth, B.G. (2001). *Chaos, Solitons and Fractals* 12, pp.613–616.
[405] Sidharth, B.G. (2001). *Chaos, Solitons and Fractals* 12, pp. 1371–1373.
[406] Davidson, M. (1978). *J.Math.Phys.* 19, (9) pp.1975–1999.
[407] Agnese, A.G. and Festa, R. (1997). *Phys.Lett. A.* 227, pp.165–171.
[408] Carneiro, S. (1998). *Found.Phys.Lett.* 11, (1), pp.95ff.
[409] Rae, A.I.M. (1986). *Quantum Mechanics* (IOP Publishing, Bristol), pp.222ff.
[410] Sidharth, B.G. *Feynman's Path Integral and Bohmian Paths* to appear in *New Advances in Physics.*
[411] P.J.E. Peebles. (1980). *The Large Scale Structure of the Universe* (Princeton University Press, Princeton).
[412] J.M.J. Geller. (1989). *Bubbles, voids and bumps in time: the new cosmology* Cornell, J. (ed.) (Cambridge University Press, Cambridge).
[413] Rubin, V.C. (1989). *Bubbles, voids and bumps in time: the new cosmology* Cornell, J. (ed.) (Cambridge University Press, Cambridge).
[414] Longair, M. (1989). *The New Physics* Davies, P. (ed.) (Cambridge University Press, Cambridge).
[415] Sidharth, B.G. and Popova, A.D. (1997). *Nonlinear World* pp.4.
[416] Pauling, L. and Wilson, E.B. (1935). *Quantum Mechanics* (McGraw Hill, Auckland).
[417] Mashhoon, B., Ho Jung Paik. and Clifford M. Will. (1989). *Physical Review D* **39**, (10), pp.2825–2838.
[418] Ruggiero, M.L. and Tartaglia, A. *Grivitomagnetic Effects, gr-qc/0207065.*
[419] Nottale, L. (2001). *Chaos, Solitons and Fractals* 12, (9), pp.1577ff.
[420] Evans, M.W. (1997). *Origin, Observation and Consequences of the* $B^{(3)}$ *Field* in *The Present Status of the Quantum Theory of Light* Jeffers, S. et al. (eds.) (Kluwer Academic Publishers, Netherlands) pp.117–125 and

several other references therein.
[421] Salhofer, Hans H. (1993). *Essays on the Formal Aspects of Electromagnetic Theory* Lakhtakia, A. (ed.) (World Scientific, Singapore), pp.268ff.
[422] Klauder, J.R. (1979). *'Bosons Without Bosons' in Quantum Theory and the Structures of Time and Space* **Vol.3** Castell, L., Van Weiizsecker, C.F. and Carl Hanser (eds.) (Verlag, Munchen).
[423] Sidharth, B.G. *arXiv/Physics 0605126.*
[424] Sidharth, B.G. (2006). *Chaos, Solitons and Fractals* 30, pp.463–469.
[425] Godel, K. (1949). *Rev.Mod.Phys.* 21, pp.447ff.
[426] Ralston, J. and Nodland, B. (1999). *Phys.Rev.Lett.* 16, pp.3043ff.
[427] Kogut, A. et al. (1997). *Phys.Rev.D.* 55, pp.1901ff.
[428] Sidharth, B.G. *arXiv/Physics 0606345.*
[429] Sachs, M. (1999). *Il Nuovo Cimento* Vol.114 B, No.2, February 1999, pp.123–126.
[430] Staruszkiewicz, A.J. (1994). *Quantum Coherence and Reality* (Y. Aharanov, Fetschrift) Anandan, J. and Sifko, J.L. (eds.) (World Scientific, Singapore), pp.90–94.
[431] Olive, D.I. (1996). *Nuc.Phys. B* (Proc.Suppl.) 46, pp.1–15.
[432] Coles, P. and Ellis, G.F.R. (1997). *Is the Universe Open or Closed?* (Cambridge University Press, Cambridge).
[433] Sidharth, B.G. (1999). *Proceedings of the Eighth Marcell Grossmann Meeting on General Relativity* Piran, T. (ed.) (World Scientific, Singapore) pp.481.
[434] Sidharth, B.G. (2002). *Chaos, Solitons and Fractals*, 13, 4, pp.617–620.
[435] Sridhar, R. (1994). *Perspectives in Theoretical Nuclear Physics* Srinivasa Rao, K. and Satpathy, L. (eds.) (Wiley Eastern, New Delhi), pp.226ff.
[436] Lieb, E.H. and Mattis, D.C. *Mathematical Physics in one dimension* (Academic Press, New York).
[437] Schonhammer, K. and Meden, V. (1996). *Am.J.Phys.* 64(9), pp.1168–1176.
[438] Bjorken, J.D. and Drell, S.D. (1965). *Relativistic Quantum Fields* (McGraw-Hill Inc., New York), pp.44ff.
[439] Whittaker, E.T. and Watson, G.N. (1962). *A Course of Modern Analysis* (Cambridge University Press), pp383.
[440] Morse, P.M. and Feschbach, H. (1953). *Methods of Theoretical Physics* (II) (Mc-Graw Hill Book Co., New York), pp1323.
[441] Ibach, H. and Luth, H. (1991). *Solid State Physics* (Narosa Publishing House, New Delhi).
[442] Sidharth, B.G. *Neutrino Mass and an Ever Expanding Universe (An Irreverent Perspective), Proceedings of Second International Symposium, "Frontiers of Fundamental Physics"* (Universities Press, Hyderabad) (in press).
[443] Sidharth, B.G. *A Note on Degenerate Bosons*, *xxx.lanl.gov/ quant-ph/9506002.*
[444] Sidharth, B.G. (1995). *Mono Energetic Fermions*, *CAMCS TR 95-04-07b.*
[445] Sidharth, B.G. *xxx.lanl.gov/phys/0008063*; (2001). *Chaos, Solitons and Fractals*, 12, pp.2475–2480.
[446] Sidharth, B.G. (2001). *Chaos, Solitons and Fractals* 12, pp.1563–1564.

[447] Sidharth, B.G. (1997). *Bull.Astr.Soc.India.* 25, pp.485–488.
[448] Schriffer, P. and Osheroff, D.D. (1995). *Rev.Mod.Phys.*, 67 (2).
[449] Delaney, P., Choi, H.J., Ihm, J., Louie, S.G. and Cohen, M.L. (1998) *Nature*, 391.
[450] Dresselhaus, M.S. (1998). *Nature*, 391.
[451] Wildoer, W.G., Liesbeth, C., Venema, G. Andrew, Rinzler, E., Richard, Smalley and Cwoes Dekker. *Nature*, 391.
[452] Odom Teri Wang., Huang Jin-Lin., Philip Kim and Charles M. Lieber. (1998) *Nature*, 391. Also Sidharth, B.G. (1999). *Low Dimensional Electrons*, in *Solid State Physics* Mukhopadhyay, R. et.al. (eds.) 41, (Universities Press, Hyderabad), pp.331.
[453] Donnelly, R.J. (1991). *Quantized Vortices in Helium II* (Cambridge University Press, Cambridge) pp.1–41.
[454] Zeilik, M. and Smith, E. (1987). *Introductory Astronomy and Astrophysics* (Saunders College Publishing, New York).
[455] Sidharth, B.G. (2003). *Journal of Indian Geophysics Union* 7 (4).

Index